Contents

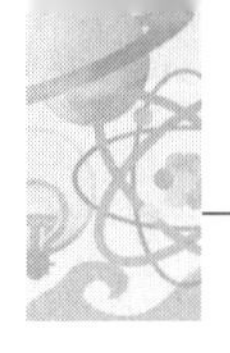

Introduction

About this guide

This unit guide covers essential material for the OCR A2 Physics A specification. It contains material needed for the effective study of **Unit G485: Fields, Particles and Frontiers of Physics**. This unit is assessed by a written examination lasting 1 hour 45 minutes. The aim of the guide is to help you *understand* the physics, so that you can effectively revise and prepare for the examination.

The **Content Guidance** section is based on the structure of the specification. The same headings are used and the sub-headings closely follow the learning outcomes of the specification. This guide is not intended to be a detailed textbook and therefore does not contain every single fact that you need to know. The focus is on understanding the principles and definitions so that you can successfully tackle the varied questions in the examination. This guide uses worked examples to illustrate good practice and offers an examiner's perspective on how you can improve your answers.

The **Question and Answer** section shows the type of questions you can expect in the unit examination. The questions are illustrated with answers given by a typical C-grade candidate and an A-grade candidate. The answers are followed by comments from the examiner, explaining why the marks were awarded or lost. Common errors made by candidates are also highlighted, so that you do not make the same mistakes. Candidates frequently lose marks for careless mistakes, incomplete answers, muddled presentation and illegible handwriting.

Using this guide

This guide can be used throughout the course and not just at revision time. The content of the guide is set out in the order of the learning outcomes of the specification, so that you can use it:

- to check your notes
- as a reference for homework and internal tests
- to revise in manageable sections
- to identify your strengths and weaknesses
- to check that you have covered the specification completely
- to learn the definitions expected by the examiners
- to improve the quality of your answers
- to familiarise yourself with the range of questions expected in examinations
- to improve your confidence in applying physics

STUDENT UNIT GUIDE

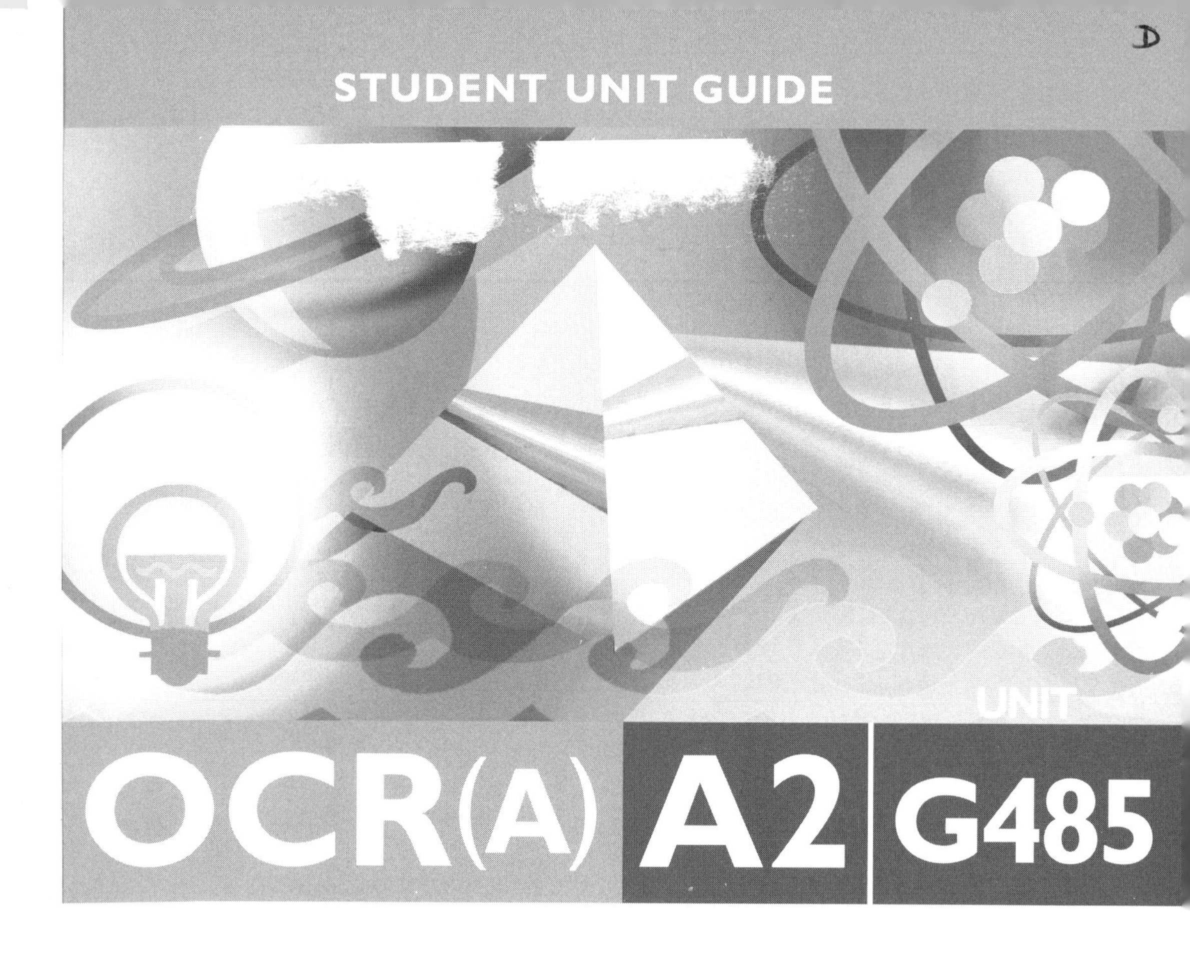

OCR(A) A2 UNIT G485

Physics

Fields, Particles and Frontiers of Physics

Gurinder Chadha

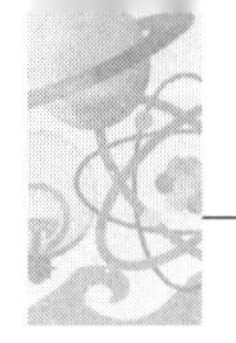

Philip Allan Updates, an imprint of Hodder Education, an Hachette UK company, Market Place, Deddington, Oxfordshire OX15 0SE

Orders

Bookpoint Ltd, 130 Milton Park, Abingdon, Oxfordshire OX14 4SB
tel: 01235 827720
fax: 01235 400454
e-mail: uk.orders@bookpoint.co.uk
Lines are open 9.00 a.m.–5.00 p.m., Monday to Saturday, with a 24-hour message answering service. You can also order through the Philip Allan Updates website: www.philipallan.co.uk

ISBN 978-0-340-95810-0

First printed 2009
Impression number 5 4 3 2 1
Year 2014 2013 2012 2011 2010 2009

Typeset by Pantek Arts Ltd, Maidstone
Printed by MPG Books, Bodmin

Revision

Planning study and revision is essential. You can avoid disappointment and anxiety by organising a revision plan well before any pending examination. You cannot cram a year's physics into a couple of days. Here are some points worth mentioning when you are revising for an examination:

- Always start with a topic that you find easy. This will boost your confidence.
- Learn all the equations listed in the specification.
- Learn the definitions and laws in the specification thoroughly. Examiners expect perfection with definitions and laws and tend to give no marks for using wrong equations.
- Ask your teacher to explain any words, laws or definitions that do not make sense.
- Make your revision active by writing out the equations, laws and definitions, drawing diagrams and doing calculations.
- Write a brief summary of the topic.
- When revising, make sure you have your notes, the specification and a reference book available to complement this guide.
- Make good use of the specification. Use a highlighter pen to identify the topics that you have covered.
- Do not revise for long periods. When you get tired and irritable, you cannot produce good quality work.
- Do not leave your revision to the last moment. Plan out a strategy spread over many weeks before the examination. Work hard during the day and learn to relax when needed.

Key skills

In examinations, candidates lose many marks because of their inability to apply some basic skills.

Calculator work

Some common mistakes are highlighted below.

- If you put $\frac{3+4}{0.5}$ into your calculator as [3] [+] [4] [÷] [0.5], you will get 11. It should be [(] [3] [+] [4] [)] [÷] [0.5], which gives the correct answer of 14.
- If you put $\frac{4}{10 \times 2}$ into your calculator as [4] [÷] [10] [×] [2], you will get 0.8. It should be [4] [÷] [(] [10] [×] [2] [)] or [4] [÷] [10] [÷] [2], which gives the correct answer of 0.2.
- If you put 3.0×10^8 into your calculator as [3.0] [×] [10] [EXP] [8], you will get 3.0×10^9. It should be [3.0] [EXP] [8]. (On some calculators, the 'exponent' button is 10^x.)
- Don't forget to insert the minus sign for powers of 10, when needed.
- Don't forget to put in squares, when they are needed.
- Take care when using lg (log to the base 10) and ln (log to the base e); make sure you are using the correct one.

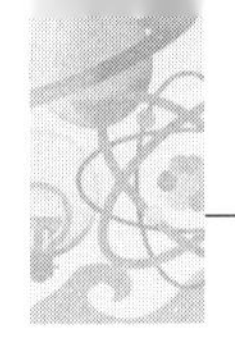

Numerical work

Some of the common mistakes are highlighted below.

- An inappropriate number of significant figures are used. If the data in a question are given to two significant figures, then your answer must also be given to two significant figures. Examiners tend not to penalise an answer with too many significant figures. In fact, candidates lose more marks by using too few significant figures.
- Rounding up an answer during a long calculation can lead to an incorrect answer. It is best to keep all the significant figures on your calculator as you progress through a calculation.
- Decimal places and significant figures may be confused. The number 0.0254 has three significant figures, though it is written to four decimal places. If you use standard form, 2.54×10^{-2} clearly has three significant figures.
- Answer are not checked to see if they are sensible. For example, you may have determined the mass of a car to be –600 kg, but you should realise that a negative mass is ridiculous. This negative answer should act as a prompt to check for an error in the calculation.

Algebraic work

Candidates lose too many marks by not being able to complete some simple algebraic operations. Here are some rules worth knowing:

- $a = b + c$ hence $b = a - c$
- $a = bc$ hence $b = \frac{a}{c}$
- $a = \frac{b}{c}$ hence $b = ac$ and $c = \frac{b}{a}$
- $a^2 = b$ hence $a = \sqrt{b}$
- $y = e^x$ hence $x = \ln(y)$

Descriptive work

In examinations, candidates tend to gain more marks from mathematical questions than questions requiring descriptive answers. Descriptive questions are often badly answered for the following reasons:

- The answer is of an inappropriate length. Keep an eye on the number of marks available for a written answer. A short answer would imply that you have missed some vital marking points and long answers can mean that either you have wasted your time or you have repeated your reasoning.
- A lack of physics in the answer. You are expected to state and use physics principles and vocabulary when answering questions requiring extended writing.
- The structure of the written answer is poor. Try not to ramble. Think carefully about the physics you want to put down on the paper.
- Poor sentence construction and bad spellings. Some questions in the examination paper have marks reserved for quality of written communication. You need to be extra careful when answering these questions.

- The question is misinterpreted. Read the question carefully and only start writing the answer once you are sure that you have understood what the question wants you to do.

Answering examination questions

Examiners do not set questions to trick you. They simply want you to demonstrate your knowledge and understanding of physics. Examiners spend many hours discussing the wording of questions and their aim is to give you the necessary information succinctly, so that you do not have to waste time deciphering the question.

When answering a question, you are expected to present your ideas logically. In an extended writing question, present your ideas in clear steps that show good use of your knowledge of physics. In a calculation, you should clearly show the following:

- correct equation
- correct substitution in the right units
- correct algebraic manipulation
- correct answer with the units

Do not waste time in the examination. You are expected on average to secure 1 mark each minute. The Unit G485 paper is worth 100 marks, but you are given 105 minutes. Read the question carefully before you put pen to paper. Make sure that all calculations are done in the correct units and you have taken account of all prefixes such as 'milli' and 'kilo'. Try to check your work as you go along. It is not sensible to check all your answers at the end of the paper because you will have forgotten the finer details of each question. You can, however, do a quick check for units and significant figures once you have finished the paper and have some spare time at the end of the session.

Many candidates waste too much time drawing diagrams using rulers. In an examination, most diagrams can be drawn freehand. This is particularly true for circuit diagrams. There is no point drawing a circuit diagram with the skills of a draughtsman when a freehand sketch showing all the components will do. Learn to save time in an examination.

In the Question and Answer section of this guide, there are many comments on the mistakes candidates make. As you get closer to the examination, make sure you read the examiner's comments at the end of each question.

Assessing 'stretch and challenge' and synopticity

Some of the A2 exam questions, referred to as 'stretch and challenge' questions, are demanding. These questions are devised for candidates hoping to attain the new A* grade. Such questions may require:

- connections to be made between different topics in physics
- open-ended responses
- extended writing
- greater mathematical competence

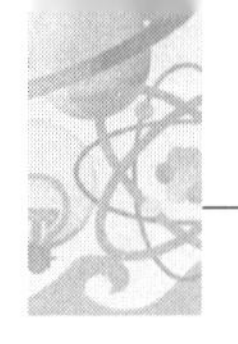

All A2 exam papers will also assess your synoptic understanding of physics. In your Unit G485 exam paper, a small number of the marks will be reserved for the application of ideas from the other units: Mechanics (G481), Electrons, Waves and Photons (G482) and The Newtonian World (G484).

Glossary of command words used in examinations

The list below shows the most frequently used command words and their meanings.

- **Calculate** — this is used when a numerical answer is required. Show all your working and give an appropriate unit for your final answer. The number of significant figures must reflect the given data.
- **Deduce** — you have to draw conclusions using the information provided.
- **Define** — a formal statement or a word equation is required. Do not use symbols unless you are prepared to define them.
- **Describe** — this requires you to state in word, using diagrams if appropriate, the main points of the topic. The amount of description depends on the mark allocation. Bullet points are acceptable unless marks are being awarded for quality of written communication.
- **Determine** — use the information available to calculate the quantity required.
- **Estimate** — this requires a calculation in which you make sensible assumptions and use realistic values for quantities. Always check whether the final answer gives a realistic value.
- **Explain** — you have to use the correct physics vocabulary and principles. The depth of your answer depends on the number of marks available.
- **Select** — you will be given a list of the key equations. Make sure you choose the most appropriate equations to do your calculations.
- **Show** — the answer to a particular problem is given in the question and is required in a subsequent calculation. You need to show each stage of your working in order to pick up all the available marks. This is not a calculator exercise, but you can use a calculator to help you reach the correct answer.
- **Sketch** — a simple freehand drawing is required. Significant detail should be added and labelled clearly.
- **Sketch a graph** — the shape of the graph needs to be correct. You may be expected to show values of the intercept or the gradient. Axes must be fully labelled and the origin shown, if appropriate.
- **State** — you are expected to write a brief answer without any supporting or justifying statements.
- **Suggest** — there is often no single correct answer. You will be given credit for sensible reasoning based on correct physics.

Data, formulae and relationships

You will be given the following information when you take the examination for Unit G485: Fields, Particles and Frontiers of Physics.

Data

Values are given to three significant figures, except where more are useful.

Speed of light in a vacuum	c	3.00×10^{8} m s^{-1}
Permittivity of free space	ε_0	8.85×10^{-12} C^{2} N^{-1} m^{-2} (F m^{-1})
Elementary charge	e	1.60×10^{-19} C
Planck constant	h	6.63×10^{-34} J s
Gravitational constant	G	6.67×10^{-11} N m^{2} kg^{-2}
Avogadro constant	N_A	6.02×10^{23} mol^{-1}
Molar gas constant	R	8.31 J mol^{-1} K^{-1}
Boltzmann constant	k	1.38×10^{-23} J K^{-1}
Electron rest mass	m_e	9.11×10^{-31} kg
Proton rest mass	m_p	1.673×10^{-27} kg
Neutron rest mass	m_n	1.675×10^{-27} kg
Alpha particle rest mass	m_α	6.646×10^{-27} kg
Acceleration of free fall	g	9.81 m s^{-2}

Conversion factors

Unified atomic mass unit	1 u = 1.661×10^{-27} kg
Electronvolt	1 eV = 1.60×10^{-19} J
Time	1 day = 8.64×10^{4} s 1 year ≈ 3.16×10^{7} s
Distance	1 light-year ≈ 9.5×10^{15} m

Mathematical equations

Arc length = $r\theta$

Circumference of circle = $2\pi r$

Area of circle = πr^2

Curved surface area of cylinder = $2\pi rh$

Volume of cylinder = $\pi r^2 h$

Surface area of sphere = $4\pi r^2$

Volume of sphere = $\frac{4}{3}\pi r^3$

Pythagoras' theorem: $a^2 = b^2 + c^2$

For small angles: $\theta \Rightarrow \sin\theta \Rightarrow \tan\theta \Rightarrow \theta$ and $\cos\theta \Rightarrow 1$

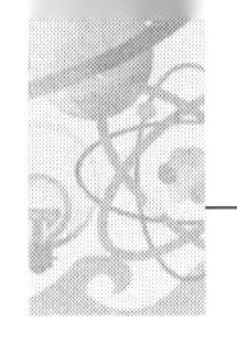

$\lg(AB) = \lg(A) + \lg(B)$

$\lg\left(\frac{A}{B}\right) = \lg(A) - \lg(B)$

$\ln(x^n) = n \ln(x)$

$\ln(e^{kx}) = kx$

Formulae and relationships

Unit G481: Mechanics

$F_x = F\cos\theta$

$F_y = F\sin\theta$

$a = \frac{\Delta v}{\Delta t}$

$v = u + at$

$s = \frac{1}{2}(u + v)t$

$s = ut + \frac{1}{2}at^2$

$v^2 = u^2 + 2as$

$F = ma$

$W = mg$

moment = Fx

torque = Fd

$\rho = \frac{m}{V}$

$p = \frac{F}{A}$

$W = Fx\cos\theta$

$E_k = \frac{1}{2}mv^2$

$E_p = mgh$

$$\text{efficiency} = \frac{\text{useful energy output}}{\text{total energy input}} \times 100\%$$

$F = kx$

$E = \frac{1}{2}kx^2$

Stress = $\frac{F}{A}$

Strain = $\frac{x}{L}$

Young modulus = $\frac{\text{stress}}{\text{strain}}$

Unit G482: Electrons, Waves and Photons

$\Delta Q = I\Delta t$

$I = Anev$

$W = VQ$

$V = IR$

$R = \frac{\rho L}{A}$

$R = R_1 + R_2 + \ldots$

$\frac{1}{R} = \frac{1}{R_1} + \frac{1}{R_2} + \ldots$

$P = VI$ $\quad P = I^2R$ $\quad P = \frac{V^2}{R}$

$W = VIt$

e.m.f. = $V + Ir$

$V_{out} = \frac{R_2}{R_1 + R_2} \times V_{in}$

$v = f\lambda$

$\lambda = \frac{ax}{D}$

$d \sin\theta = n\lambda$

$E = hf$ $\quad E = \frac{hc}{\lambda}$

$hf = \phi + \text{KE}_{max}$

$\lambda = \frac{h}{mv}$

Unit G484: The Newtonian World

$F = \frac{\Delta p}{\Delta t}$

$v = \frac{2\pi r}{T}$

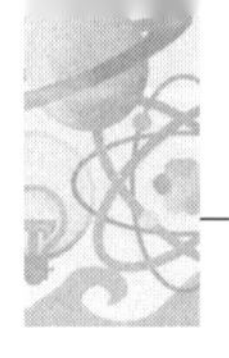

$$a = \frac{v^2}{r}$$

$$F = \frac{mv^2}{r}$$

$$F = -\frac{GMm}{r^2}$$

$$g = \frac{F}{m}$$

$$g = -\frac{GM}{r^2}$$

$$T^2 = \left(\frac{4\pi^2}{GM}\right)r^3$$

$$f = \frac{1}{T}$$

$$\omega = \frac{2\pi}{T} = 2\pi f$$

$$a = -(2\pi f)^2 x$$

$$x = A\cos(2\pi ft)$$

$$v_{max} = (2\pi f)A$$

$$E = mc\Delta\theta$$

$$pV = NkT$$

$$pv = nRT$$

$$E = \frac{3}{2}kT$$

Unit G485: Fields, Particles and Frontiers of Physics

$$E = \frac{F}{Q}$$

$$F = \frac{Qq}{4\pi\varepsilon_0 r^2}$$

$$E = \frac{Q}{4\pi\varepsilon_0 r^2}$$

$$E = \frac{V}{d}$$

$F = BIL\sin\theta$

$F = BQv$

$\phi = BA\cos\theta$

induced e.m.f. = –rate of change of magnetic flux linkage

$\frac{V_s}{V_p} = \frac{n_s}{n_p}$

$Q = VC$

$W = \frac{1}{2}QV$ $W = \frac{1}{2}CV^2$

time constant = CR

$x = x_0 e^{-\frac{t}{CR}}$

$C = C_1 + C_2 + \ldots$

$\frac{1}{C} = \frac{1}{C_1} + \frac{1}{C_2} + \ldots$

$A = \lambda N$

$A = A_0 e^{-\lambda t}$

$N = N_0 e^{-\lambda t}$

$\lambda t_{1/2} = 0.693$

$\Delta E = \Delta mc^2$

$I = I_0 e^{-\mu x}$

$Z = \rho c$

$\frac{I_r}{I_o} = \frac{(Z_2 - Z_1)^2}{(Z_2 + Z_1)^2}$

$\frac{\Delta\lambda}{\lambda} = \frac{v}{c}$

age of universe = $\frac{1}{H_0}$

$\rho_0 = \frac{3H_0^2}{8\pi G}$

This section is a student's guide to the A2 Unit G485: Fields, Particles and Frontiers of Physics. It covers all the relevant key facts, explains the essential concepts and highlights common misconceptions. The main topics are:

- Electric fields
- Magnetic fields
- Electromagnetism
- Capacitors
- The nuclear atom
- Fundamental particles
- Radioactivity
- Nuclear fission and fusion
- X-rays
- Diagnostic methods in medicine
- Ultrasound
- Structure of the universe
- The evolution of the universe

Electric fields

Electric field strength

A charged object is surrounded by an electric field. This means that if a charged particle ventures into this electric field, then it will experience a force. The magnitude of the force experienced by the charged particle depends on its charge and the strength of electric field. The electric field strength at a point in space is defined as follows:

The electric field strength *E* at a point is the force experienced per unit charge exerted on a positive charge placed at that point.

That is:

$$E = \frac{F}{Q}$$

where F is the force experienced by a positive charge of magnitude Q.

The unit for electric field strength is $\mathrm{N\,C^{-1}}$. Electric field strength is a *vector*; it has both magnitude and direction.

Electric field patterns

Electric field patterns can be mapped out using electric *field lines*. The direction of the field at a point in space shows the direction of the force experienced by a small positive charge placed at that point. This is why electric field lines always point away from positively charged objects and, conversely, point towards negatively charged objects.

It is worth remembering that all electric fields are created by fundamental particles such as electrons and protons. A single particle such as an electron or a proton produces a *radial* electric field. A neutral atom has an equal number of electrons and protons. Hence, it does not produce an overall electric field because it has zero net charge.

The diagram on p. 18 shows some electric field patterns.

Note:

- Electric field lines always have direction.
- Electric field lines never cross.
- Closely spaced electric field lines indicate greater electric field strength – a 'stronger' field.

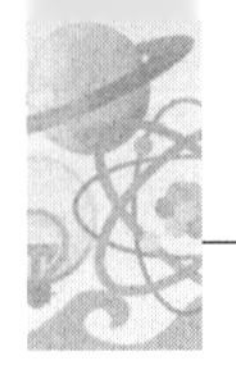

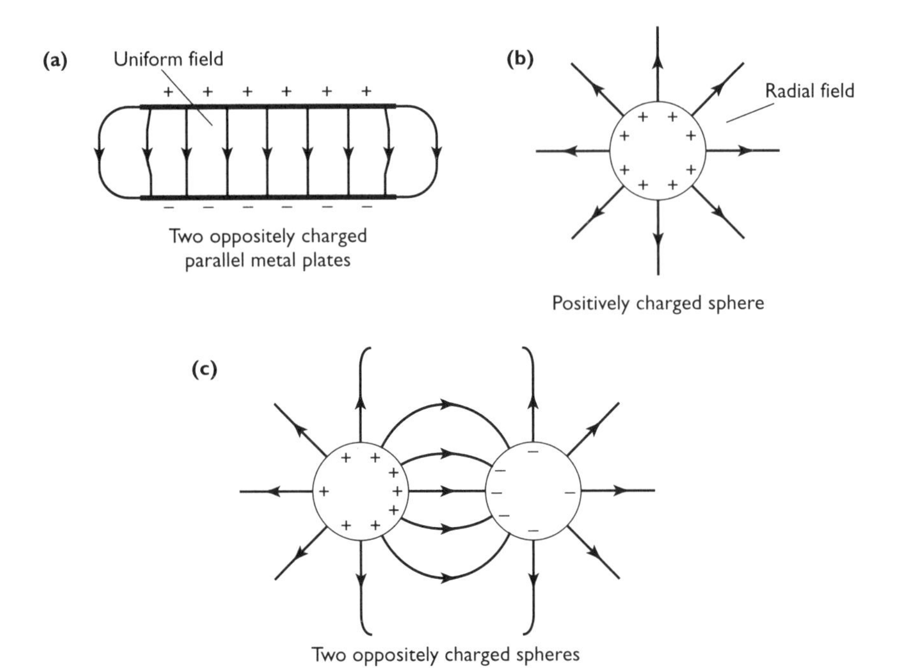

Uniform electric field

The diagram below shows two parallel plates connected to a battery.

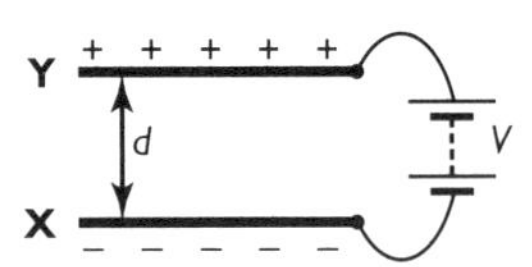

The plates are oppositely charged. The positive plates are deficient of electrons and the negative plate has an equal number of excess electrons. The electric field is *uniform* for most of the region between the plates. This means that the electric field strength is constant. How can we tell this from the electric field pattern? The electric field lines are parallel and evenly spaced (except at the edges).

The potential difference between the plates is V and the plates are separated by a distance d. Consider a positive charge Q that is moved from the negative plate X to the positive plate Y at a constant velocity.

work done on the charge = energy transformed

$F \times d = V \times Q$

Rearranging this equation gives:

$$\frac{F}{Q} = \frac{V}{d}$$

The ratio F/Q is the magnitude of the electric field strength E between the parallel plates. Therefore:

$$E = \frac{V}{d}$$

A common mistake by candidates is to quote the equation above for the definition of electric field strength. It is worth noting that the equation above only applies to charged parallel plates.

Note: the alternative unit for electric field strength is V m^{-1}.

Worked example

An electron is accelerated between two charged parallel plates. The p.d. between the plates is 450 V and the plates are separated by 1.8 cm. The electron travels from the negative plate to the positive plate. For this electron, calculate:

(a) the force it experiences due to the electric field

(b) its acceleration

(c) its final velocity at the positive plate (assume the electron's initial velocity is zero)

Answer

(a) The uniform electric field between the plates is given by $E = \frac{V}{d}$.

Hence:

$$E = \frac{V}{d} = \frac{450}{1.8 \times 10^{-2}} = 2.5 \times 10^{4}\,\text{V m}^{-1}$$

The force F on the electron is constant and is given by:

$F = EQ = 2.5 \times 10^{4} \times 1.60 \times 10^{-19} = 4.0 \times 10^{-15}\,\text{N}$

This force accelerates the electron towards the positive plate.

(b) $a = \frac{F}{m} = \frac{4.0 \times 10^{-15}}{9.11 \times 10^{-31}} = 4.39 \times 10^{15}\,\text{m s}^{-2} \approx 4.4 \times 10^{15}\,\text{m s}^{-2}$

(c) The acceleration of the electron is constant. Hence, we can use the following equation of motion to find its final velocity:

$v^2 = u^2 + 2as$ and $u = 0$

$v = \sqrt{2 \times 4.39 \times 10^{15} \times 1.8 \times 10^{-2}} = 1.26 \times 10^{7}\,\text{m s}^{-1} \approx 1.3 \times 10^{7}\,\text{m s}^{-1}$

Note: part (c) of this question is synoptic. It requires knowledge from Unit G481. However, do not forget that all the equations of motion are given in the *Data, Formulae and Relationships* booklet.

Coulomb's law

Here are two important rules of electrostatics:

- Like charges repel.
- Unlike charges attract.

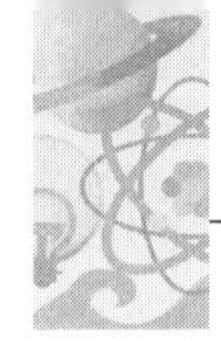

Like particles such as electrons will repel each other. The force between two unlike particles, such as a proton and an electron, will be attractive. This simple rule gives the direction of the force, but not its magnitude.

It was the French physicist Charles Coulomb who first proposed a law to describe the magnitude of the force between two charged particles. A statement of his famous law, **Coulomb's law**, is as follows:

> **Two point charges exert an electrical force on each other that is directly proportional to the product of the charges and inversely proportional to the square of the separation between them.**

The diagram below shows two charged particles (point charges) interacting.

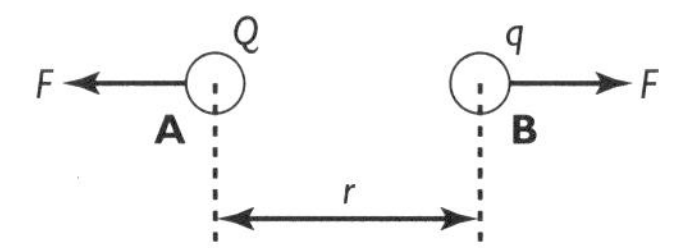

Particle A experiences a force because it lies in the electric field created by particle B and particle B experiences a force because it lies in the electric field created by particle A. According to Newton's third law, each particle experiences an *equal* but *opposite* force.

According to Coulomb's law, the magnitude of the force F between the charged particles is:

- directly proportional to the product of the charges
- inversely proportional to the square of the separation

That is:

$$F \propto Qq$$

and

$$F \propto \frac{1}{r^2}$$

The relationships above may be combined to give an equation for the force F. The equation for the force F is:

$$F = \frac{Qq}{4\pi\varepsilon_0 r^2}$$

The constant ε_0 is known as the permittivity of free space. It has an experimental value of $8.85 \times 10^{-12}\,\mathrm{F\,m^{-1}}$.

Note: you can also apply Coulomb's law to uniformly charged spheres as long as you measure the distance r from the *centres* of the spheres. It is as if the charges were concentrated at the centre of each sphere.

Worked example

Two identical spheres are charged. Each sphere has radius 2.5 cm and charge +0.30 nC. Calculate the force between two charged spheres when their *surfaces* are separated by a distance of 2.0 cm.

Answer

The force experienced by each charged sphere is given by the equation

$F = \dfrac{Qq}{4\pi\varepsilon_0 r^2}$, where r is the distance between the *centres* of the spheres.

$$Q = q = 0.30 \times 10^{-9}\,\text{C}$$

$$r = (2.0 + 2.5 + 2.5) = 7.0\,\text{cm} = 0.07\,\text{m}$$

$$F = \frac{Qq}{4\pi\varepsilon_0 r^2} = \frac{(0.30 \times 10^{-9})^2}{4\pi \times 8.85 \times 10^{-12} \times (0.07)^2}$$

$$F = 1.65 \times 10^{-7}\,\text{N} \approx 1.7 \times 10^{-7}\,\text{N}$$

Note: this worked example shows how important it is to understand that the distance r must be a centre-to-centre distance for the spheres. Examiners would deduct marks if 2.0 cm was used for the separation r.

Electric field strength for radial field

What is the electric field strength E at a distance r from a point charge or the centre of a uniformly charged sphere?

By definition, the electric field strength is the *'force per unit positive charge'*.

$$E = \frac{\text{force}}{\text{charge}}$$

$$E = \frac{Qq}{4\pi\varepsilon_0 r^2} \div q$$

Therefore:

$$E = \frac{Q}{4\pi\varepsilon_0 r^2}$$

The electric field strength is inversely proportional to the square of the distance; doubling the distance will decrease the field strength by a factor of 4 etc. A common mistake made by candidates is to use the equation $E = \dfrac{V}{d}$ for a charged particle; this equation only applies to *uniform* electric fields.

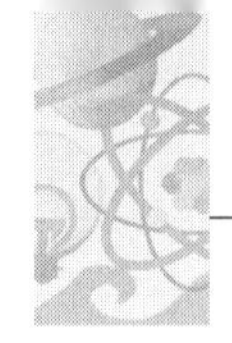

Comparing electric and gravitational fields

Both Coulomb's law and Newton's law of gravitation have one important similarity — the force between objects obeys an inverse square law with distance. The section below shows the similarities and differences between electrical fields and gravitational fields.

Similarities

- Both electric and gravitational fields are to do with 'action at a distance'.
- The field strengths for both obey an inverse square law with distance; $E \propto \frac{1}{r^2}$ and $g \propto \frac{1}{r^2}$.
- A point mass and a point charge both produce radial fields.
- The field strength is defined as 'force per unit mass or positive charge'.

Differences

- An electric field is created by *charge*, whereas a gravitational field is created by *mass*.
- Electric fields can be attractive or repulsive. Gravitational fields are always attractive.

Magnetic fields

Magnetic field patterns

A bar magnet produces a magnetic field in the space around it. Another magnet brought into this magnetic field will experience a force. Over the years, physicists have concluded that a magnetic field is created by moving charged particles. The magnetic field of a bar magnet is the result of tiny magnetic fields created by electrons moving around in the atoms of the material of the magnet.

Magnetic field patterns can be mapped out using magnetic field lines. The direction of the field at a point in space shows the direction of the force experienced by a 'free north pole' at that point. This is why magnetic field lines always point away from a north pole and, conversely, point towards a south pole.

The diagram below shows (a) the non-uniform magnetic field of a single bar magnet and (b) the uniform magnetic field between the opposite poles of a strong magnet.

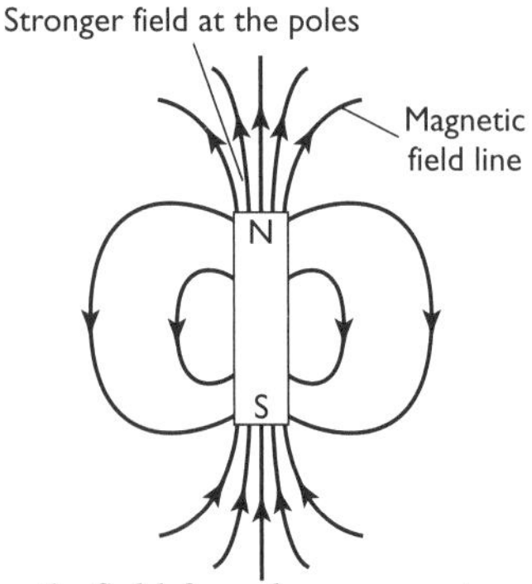

(a) Magnetic field for a bar magnet

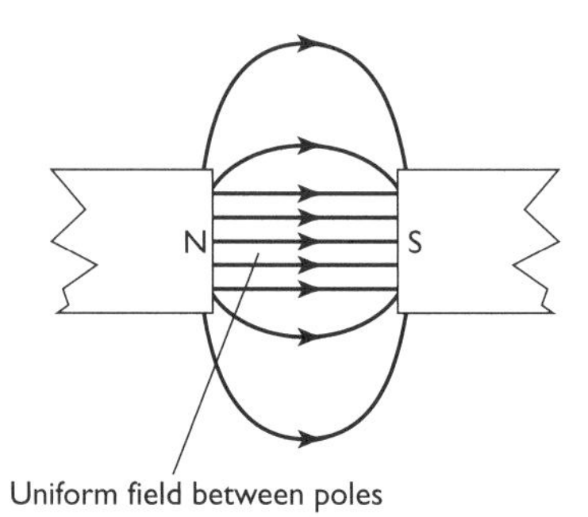

(b) Field between the poles of a magnet

Electromagnetism

Moving charges create a magnetic field. The magnetic field disappears when the charges stop moving. Hence, a magnetic field can be created by an electric current in a wire.

The diagram below shows the magnetic field patterns for (a) a long, straight current-carrying conductor and (b) a solenoid.

For a current-carrying conductor:

- the magnetic field lines are *concentric circles* round the conductor
- you can use the right-hand grip rule to determine the direction of the magnetic field from the direction of conventional current

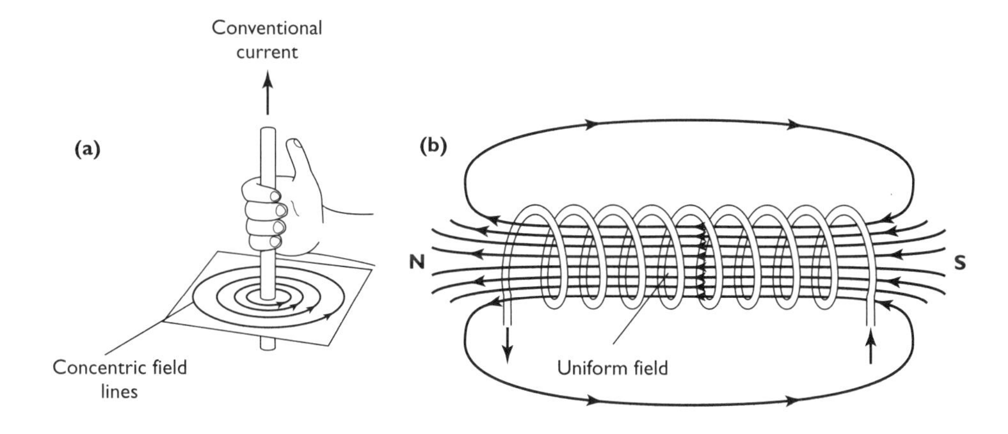

Right-hand grip rule: the thumb points in the direction of the conventional current and the direction in which the fingers wrap round the conductor gives the direction of the magnetic field.

For a solenoid:

- the magnetic field resembles that for a bar magnet
- the magnetic field is *uniform* within the core of the solenoid
- there is a north pole at one end and a south pole at the opposite end

Fleming's left-hand (motor) rule

Closely placed magnets experience a force because their magnetic fields interact.

A current-carrying wire experiences a force when placed in an external magnetic field of a magnet because their magnetic fields also interact. The direction of the force experienced by a current-carrying conductor placed at right angles to a magnetic field can be determined using Fleming's left-hand rule (see diagram on p. 24).

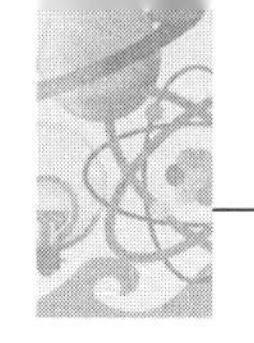

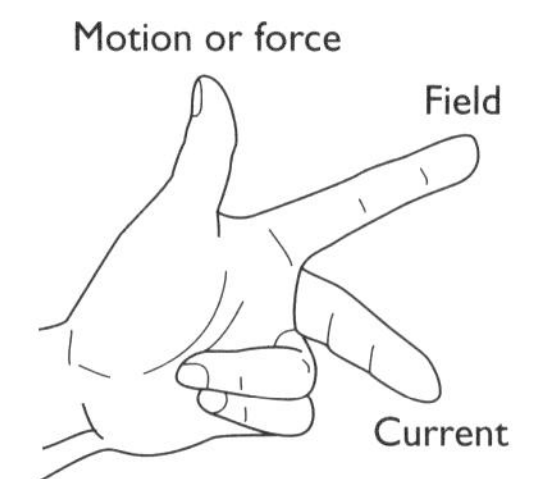

thuMb = Motion or force
First finger = Field
SeCond finger = conventional Current

Magnetic flux density

As with electric and magnetic fields, the *strength* of the magnetic field is stronger where the magnetic fields lines are close together. The correct term used for the strength of the magnetic field is magnetic flux density. This is defined in terms of the force experienced by a current-carrying wire placed at right angles to a magnetic field.

The magnetic flux density B is defined by the equation:

$$B = \frac{F}{IL}$$

where F is the force on the current-carrying wire, I is the current in the wire and L is the length of the wire in the magnetic field. The direction of the force can be determined using Fleming's left-hand rule.

The unit for magnetic flux density is the tesla, T.

The tesla is defined as follows:

The magnetic flux density is 1 T when a wire carrying a current of 1 A placed at right angles to the field experiences a force of 1 N per metre of its length.

Note: $1\,\text{T} = 1\,\text{N}\,\text{m}^{-1}\,\text{A}^{-1}$

The force on a current-carrying wire placed at 90° to an external magnetic field is given by:

$$F = BIL$$

When the current-carrying wire is placed at an angle θ to the magnetic field, then the magnitude of the force on the wire is given by:

$$F = BIL \sin\theta$$

Worked example

A thin wire has a weight of $1.6 \times 10^{-3}\,\text{N}\,\text{cm}^{-1}$. The wire is placed at right angles to the magnetic field of flux density 0.20 T. The direction of the current in the wire is such that it experiences an upward force. The current is slowly increased. Calculate the current in the wire when the force on the wire due to the magnetic field is equal to the weight of the wire.

Answer

The force acting on 1.0 cm section of the wire is equal to 1.6×10^{-3} N.

$B = 0.20$ T, $I = ?$, $L = 0.01$ m, $F = 1.6 \times 10^{-3}$ N

Since $\theta = 90°$, the force on the wire is given by the equation $F = BIL$

$$I = \frac{F}{BL} = \frac{1.6 \times 10^{-3}}{0.20 \times 0.01} = 0.80\,\text{A}$$

The current in the wire is 0.80 A when the wire becomes 'self supportive'.

Moving charges in a magnetic field

A current-carrying wire placed in a magnetic field experiences a force because each moving electron within the wire experiences a tiny force. The force F on a positive particle moving at right angles to magnetic field is given by the equation:

$$F = BQv$$

where B is the magnetic flux density, Q is the charge on the particle and v is the speed of the particle. The direction of the force is given by Fleming's left-hand rule.

For an electron, the charge Q is e, the elementary charge. Hence:

$$F = Bev$$

The diagram below shows the path of electrons as they enter a region of uniform magnetic field.

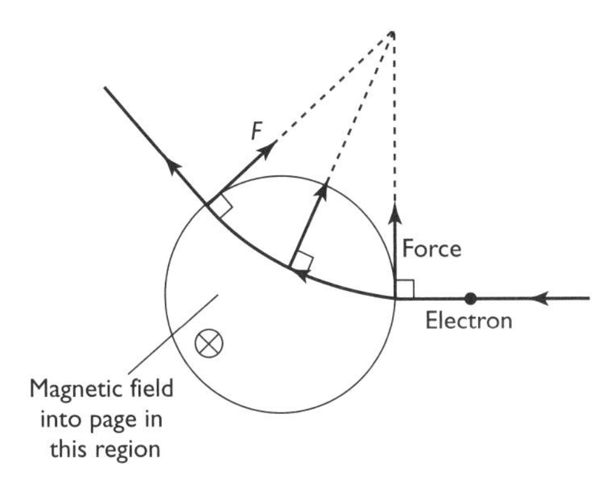

Note:

- The force F on the electron is at right angles to its velocity; hence, the path of the electron in the magnetic field will be a circle.
- The magnetic force (BQv) provides the necessary centripetal force for circular motion.
- The list of equations on p. 26 is all you need to solve most problems related to charged particles moving in a uniform magnetic field (including the spectrometer – see p. 27).

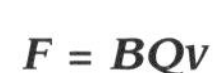

$$\boldsymbol{F = BQv} \qquad \textbf{(1)}$$

$$\boldsymbol{F = ma} \qquad \textbf{(2)}$$

$$\boldsymbol{F = \frac{mv^2}{r}} \qquad \textbf{(3)}$$

$$\boldsymbol{v = \frac{2\pi r}{T}} \qquad \textbf{(4)}$$

The following worked example shows how these equations are useful.

Worked example

An electron travelling at $7.5 \times 10^6\,\text{m s}^{-1}$ enters a region of uniform magnetic field of flux density $60\,\mu\text{T}$. The electron is initially travelling at right angles to the magnetic field. Calculate the radius of the circle described by the electron in the magnetic field.

Answer

The force is right angles to the velocity of the electron, so the path is a circle in the region of the magnetic field.

The magnetic force 'Bev' provides the centripetal force on the electron. Therefore

$$Bev = \frac{mv^2}{r}$$

$$r = \frac{mv}{Be}$$

$$\text{radius} = \frac{9.11 \times 10^{-31} \times 7.5 \times 10^{6}}{60 \times 10^{-6} \times 1.60 \times 10^{-19}} = 0.712\,\text{m} \approx 0.71\,\text{m}$$

There is an alternative method: calculate the force on the electron using $F = Bev$ and then use $F = \frac{mv^2}{r}$ to find the radius.

A velocity selector

The diagram below shows a velocity selector.

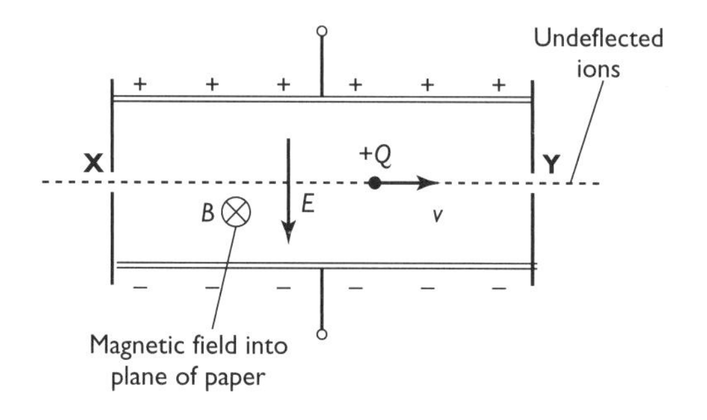

A velocity selector consists of two oppositely charged plates. These provide a uniform electric field. A uniform magnetic field is also applied at right angles to the electric field. Charged particles of different speeds enter at X and only those with a particular speed v will travel straight through the slit Y. The electric and magnetic fields are such that the electrical force on the charged particle is opposite in direction to the magnetic force.

For the charged particles travelling in a straight line between X and Y, we have:

electric force = magnetic force

$$EQ = BQv$$

The charge Q on the particle cancels out, therefore:

$$v = \frac{E}{B}$$

Charged particles with speeds other than given by the equation will not travel in a straight line and will miss the slit Y.

Mass spectrometer

A mass spectrometer is used to determine the masses of charged ions and their relative abundances. The diagram below shows a typical spectrometer.

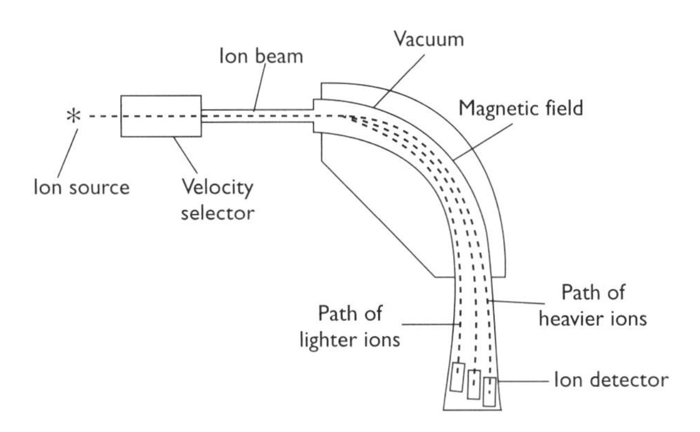

For a mass spectrometer:

- a velocity selector is used to isolate ions of a particular speed
- a uniform magnetic field is used to deflect the charged ions in a circular path in an evacuated chamber
- a movable detector, or photographic film, is used to determine the radius of the path and the relative abundance of the ions
- the radius r of the path in the magnetic field region is given by the equation $r = \frac{mv}{Be}$ (note that the radius r is directly proportional to the mass of the charged ion)

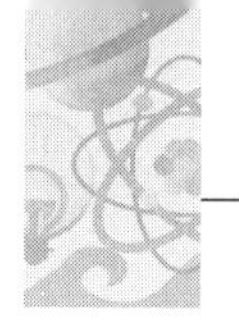

Modern mass spectrometers can measure the masses of ions to great precision. For example, the mass of a helium ion is 6.64513×10^{-27} kg.

Electromagnetism

Important definitions

This section deals with many quantities that sound similar but have very different meanings. The three important quantities are:

- magnetic flux density B
- magnetic flux ϕ
- magnetic flux linkage

Magnetic flux ϕ is defined by the following word equation:

magnetic flux = magnetic flux density × cross-sectional area normal to the magnetic field

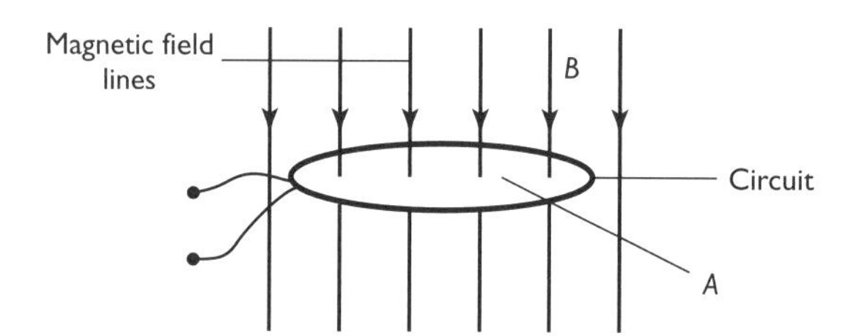

Therefore:

$$\phi = BA$$

If the magnetic field is not normal to a circuit, then we have to determine the component of the magnetic flux density at right angles to the area of the circuit. Hence, for a magnetic field making an angle θ to the normal of the coil, we have:

$$\phi = BA\cos\theta$$

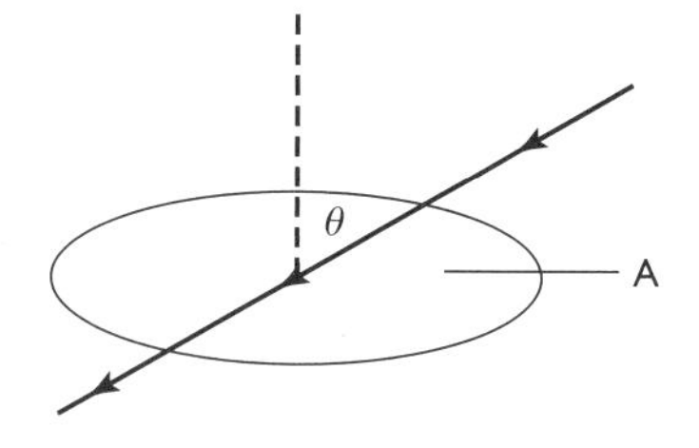

The unit for magnetic flux is the weber (Wb). The weber is defined as:

A magnetic field of flux density 1 T passing normally through an area of 1 m² produces a magnetic flux equal to 1 Wb

Note: $1\,\text{Wb} = 1\,\text{T m}^2$

Magnetic flux *linkage* is defined by the following word equation:

magnetic flux linkage = number of turns × magnetic flux

Therefore:

magnetic flux linkage = $N\phi$

The unit for magnetic flux linkage is also the weber (Wb).

Faraday's law of electromagnetic induction

An e.m.f. is induced in a circuit whenever there is a change in the magnetic flux linkage. The *magnitude* of the induced e.m.f. in a circuit can be determined using Faraday's law of electromagnetic induction:

The magnitude of the induced e.m.f. is equal to the rate of change of flux linkage

Therefore, the magnitude of induced e.m.f. E is given by:

$$E = \frac{\Delta(N\phi)}{\Delta t}$$

There are three ways in which an e.m.f. may be induced in a circuit:

- change the magnetic flux density B (e.g. move a coil closer to the pole of a bar magnet)
- change the area A of the circuit (e.g. move a straight wire at right angles to the magnetic field)
- change the angle θ (e.g. rotate the coil, as in a generator)

The direction of the induced e.m.f. in a circuit is governed by the principle of *conservation of energy* and this in turn, is stated by Lenz's law:

The direction of induced e.m.f. or the current is such as to oppose the change that is producing it

A complete equation for the induced e.m.f. in a circuit has a minus sign because of Lenz's law. Therefore:

$$E = -\frac{\Delta(N\phi)}{\Delta t}$$

Worked example

A flat coil of 800 turns has cross-sectional area $7.0 \times 10^{-4}\,\text{m}^2$ and is connected to the terminals of an ammeter. The total resistance of the coil and the ammeter is $0.30\,\Omega$. The plane of the coil is at right angles to a magnetic field of flux density 0.12 T. The coil is removed from the magnetic field in 50 ms. Calculate the average induced e.m.f. across the coil and the average current shown by the ammeter.

Answer

$$E = -\frac{\Delta(N\phi)}{\Delta t} = -\frac{\Delta(NBA)}{\Delta t}$$

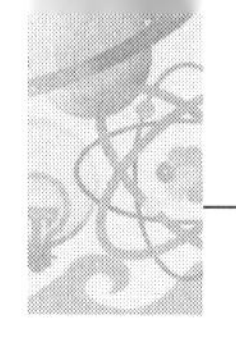

The final flux linkage is zero when the coil is withdrawn from the magnetic field.

$$E = -\frac{0 - (800 \times 0.12 \times 7.0 \times 10^{-4})}{50 \times 10^{-3}}$$

e.m.f. = 1.34 V (magnitude only)

$$\text{average current} = \frac{\text{e.m.f.}}{\text{resistance}} = \frac{1.34}{0.30} \approx 4.5\text{ A}$$

The average current in the coil is 4.5 A.

Induced e.m.f. across the ends of a wire

A wire of length L is placed at right angles to a uniform magnetic field of flux density B. It is moved at a constant speed v. An e.m.f. will be induced across the ends of the wire (see the diagram below).

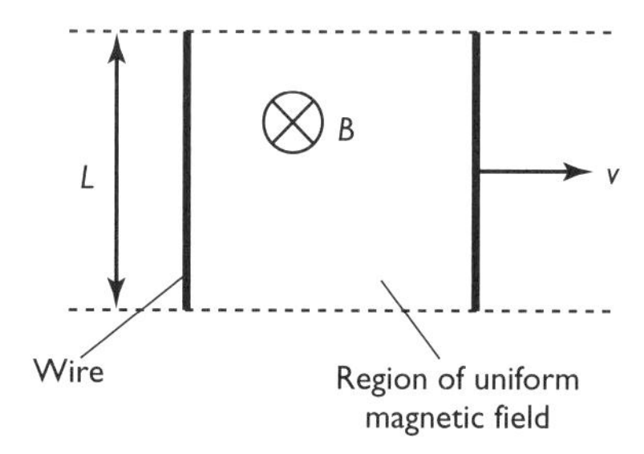

The magnitude of the induced e.m.f. E can be determined as follows:

$$E = \frac{\Delta(NBA)}{\Delta t} = NB \times \frac{\Delta A}{\Delta t}$$

The 'circuit' has '1 turn' and the rate of change of area = Lv. Therefore

$$E = NB \times (Lv)$$

$$E = BvL$$

Simple generator

A simple generator consists of a rectangular coil rotating at a constant frequency in a uniform magnetic field. The rotation of the coil induces an e.m.f. across the ends of the coil; some of coil's kinetic energy is transformed into electrical energy.

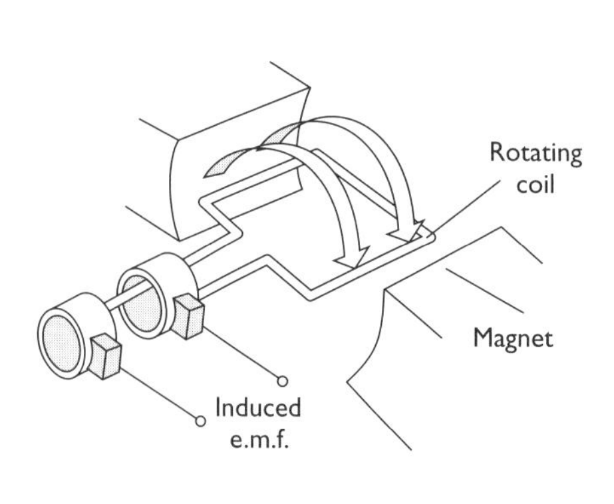

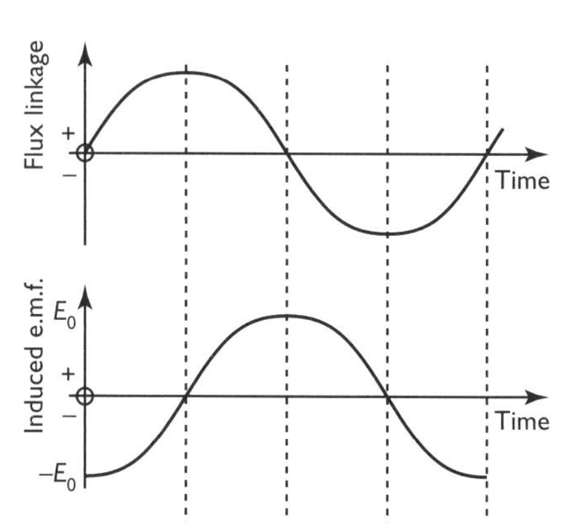

The diagram on p. 30 shows a simple a.c. generator. Also shown is the variation of the magnetic flux linkage with time and the corresponding variation of the e.m.f. across the ends of the rotating coil. The e.m.f. induced across the ends of the coil is equal to the *'rate of change of flux linkage'*, hence the magnitude of the induced e.m.f. is equal to the *gradient* of flux linkage against time graph.

Note: the induced e.m.f. is zero when the flux linkage is a maximum. The induced e.m.f. is a maximum when the flux linkage is zero.

The 'peak' e.m.f. E_0 (the maximum induced e.m.f.) is directly proportional to:

- the frequency f of the rotating coil
- the magnetic flux density B
- the number of turns N on the coil
- the cross-sectional area A of the coil

Transformers

The diagram below shows a simple transformer.

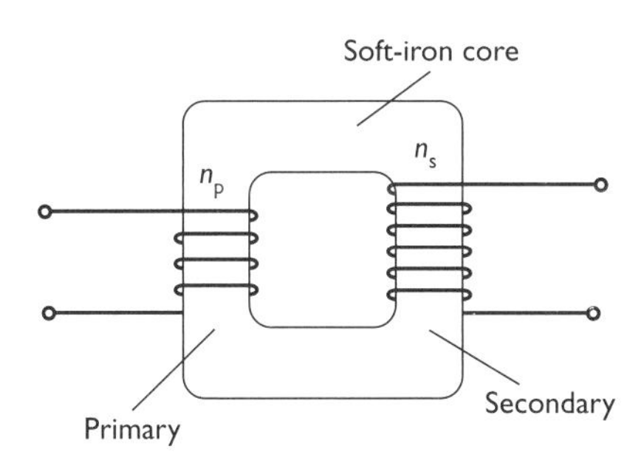

A transformer consists of two coils mounted on an iron core. The primary coil is the 'input' coil and is connected to a source of alternating current. The secondary coil is the 'output' coil. The iron core is used to ensure that all the magnetic flux created by the primary coil links the secondary coil.

How does a transformer function?

The alternating current of the primary coil produces a *changing* magnetic field. This changing magnetic field links the secondary coil. There is change in the rate of flux linkage in the secondary coil. An e.m.f. is therefore induced in the secondary coil.

We can use the following turn-ratio equation to determine the size of the induced e.m.f. at the secondary coil:

$$\frac{V_s}{V_p} = \frac{n_s}{n_p}$$

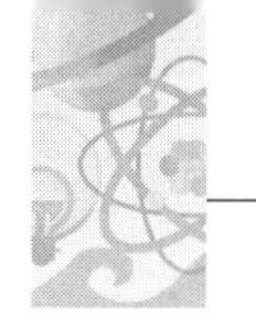

V_s is the output voltage at the secondary coil, V_p is the input voltage at the primary coil, n_s is the number of turns on the secondary coil and n_p is the number of turns on the primary coil.

- For a *step-up* transformer, $n_s > n_p$ and the output voltage is greater than the input voltage.
- For a *step-down* transformer, $n_s < n_p$ and the output voltage is smaller than the input voltage.

For a 100% efficient transformer:

input power = output power

$$\boldsymbol{V_p I_p = V_s I_s}$$

where I_p and I_s are the currents in the primary coil and secondary coil respectively.

Capacitors

Capacitance

A capacitor consists of two metal plates separated by an insulator (air, ceramic, mica etc.). When connected to a source of e.m.f., the plates acquire *equal* but *opposite* charges. The positive plate loses electrons and the negative plate gains an equal number of electrons.

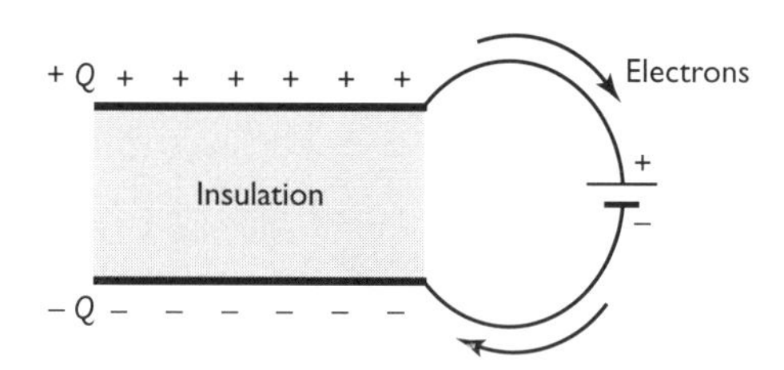

The magnitude of the charge Q on one of the plates is directly proportional to the p.d. V across capacitor. That is:

$$Q \propto V$$

For a given capacitor, we have

$$\boldsymbol{Q = VC}$$

where C is the capacitance of the capacitor.

The capacitance of a capacitor is defined by the following word equation:

$$\textbf{capacitance} = \frac{\textbf{charge}}{\textbf{p.d.}}$$

The unit for capacitance is the farad (F). This is defined as follows:

A capacitor has a capacitance of 1 F when it can store a charge of 1 coulomb per unit volt.

Note: $1\,F = 1\,C\,V^{-1}$

Worked example

The p.d. across a 1000 μF capacitor is changed from 2.0 V to 6.0 V. Calculate the charge in mC gained by each capacitor plate and the number of electrons gained by the negative plate.

Answer

$Q = VC$

initial charge = $VC = 2.0 \times 1000 = 2000\,\mu C$

final charge = $VC = 6.0 \times 1000 = 6000\,\mu C$

charge gained = $6000 - 2000 = 4000\,\mu C$

The charge on each electron is $e = 1.60 \times 10^{-19}\,C$. Hence the number N of electrons gained by the negative plates is:

$$N = \frac{4000 \times 10^{-6}}{1.60 \times 10^{-19}} = 2.5 \times 10^{16}$$

Capacitors in series

The diagram below shows capacitors of capacitances C_1, C_2 and C_3 connected in series.

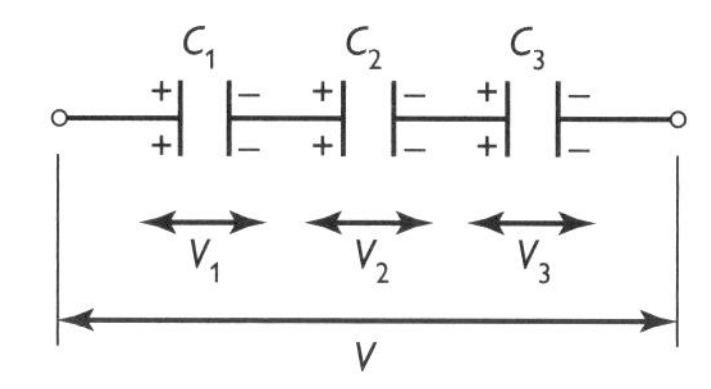

The arrangement above can be replaced by a single capacitor of capacitance C. For this series arrangement of capacitors, the following rules apply:

- The *charge* on each capacitor is the *same*.
- The total capacitance C is given the equation $\frac{1}{C} = \frac{1}{C_1} + \frac{1}{C_2} + \ldots$.
- The total p.d. across the combination is the sum of individual p.d.s across the capacitors, that is, $V = V_1 + V_2 + V_3 + \ldots$.

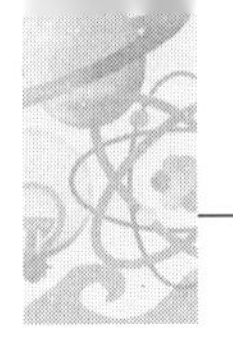

Capacitors in parallel

The diagram below shows capacitors of capacitances C_1, C_2 and C_3 connected in parallel.

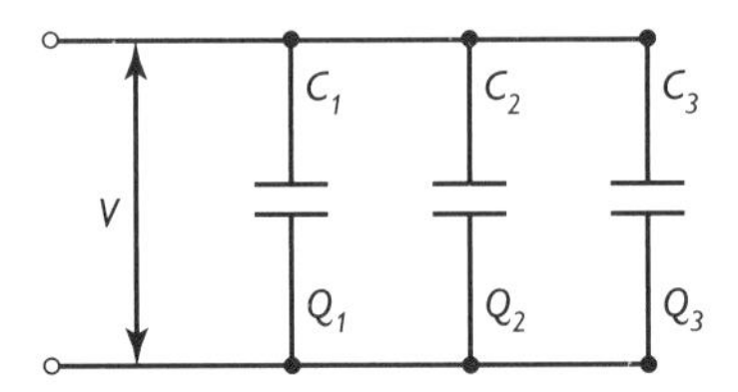

The arrangement above can be replaced by a single capacitor of capacitance C. For this parallel arrangement of capacitors, the following rules apply:

- The total charge Q stored by the arrangement is the sum of the charges stored in the individual capacitors, that is $Q = Q_1 + Q_2 + Q_3 + \dots$.
- The total capacitance C is given the equation $C = C_1 + C_2 + C_3 + \dots$.
- The *voltage* (*p.d*). across each capacitor is the *same*.

Energy stored by capacitor

Consider a capacitor plate that is charged negative. In order to increase the number of electrons on this plate, work has to be done *against* the repulsive forces. Since work done is equal to energy, this implies that a capacitor will store electrical energy.

The diagram below shows a graph of voltage against charge for a capacitor.

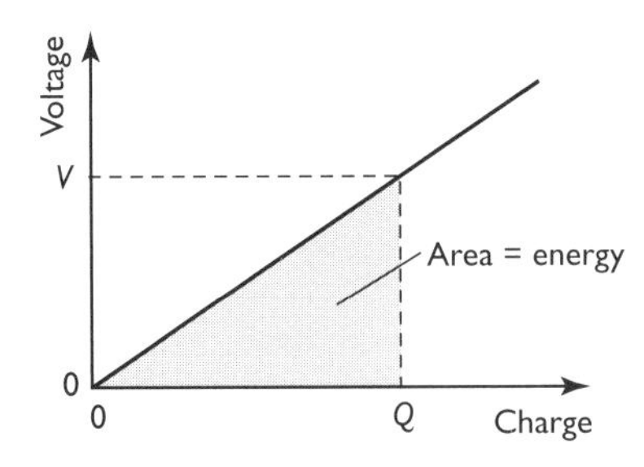

What information can we extract from such a graph?

- The reciprocal of the gradient is equal to capacitance C of the capacitor; that is gradient = $1/C$.
- The *area* under the graph is equal to the *energy* (or work done).

For a capacitor with charge Q and voltage V, the energy stored E by the capacitor is the area of a triangle of 'height' V and 'base' Q.

Therefore:

$$E = \frac{1}{2}QV$$

Since $Q = VC$, we also have alternative equations for the stored energy:

$$E = \frac{1}{2}CV^2 \text{ and } E = \frac{1}{2}\frac{Q^2}{C}$$

Worked example

A 0.02 F capacitor is charged to 9.0 V. The fully charged capacitor is discharged through a lamp in a time of 40 ms. Calculate the average power dissipated by the lamp.

Answer

$C = 0.02\,\text{F}$, $V = 9.0\,\text{V}$

$$E = \frac{1}{2}CV^2 = \frac{1}{2} \times 0.02 \times 9.0^2 = 0.81\,\text{J}$$

The power dissipated is given by power = energy/time. Therefore:

$$\text{power} = \frac{0.81}{0.040} = 20.3\,\text{W} \approx 20\,\text{W}$$

Capacitor discharging through a resistor

The diagram below shows a capacitor of capacitance C connected across a resistor of resistance R.

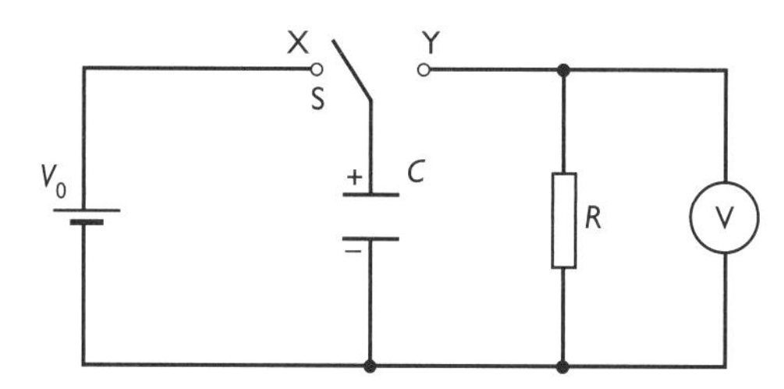

With the switch S at position X the capacitor is fully charged. The p.d. across the capacitor is V_0. When the switch is connected to position Y, the capacitor will discharge through the resistor. The capacitor is in *parallel* with the resistor, hence the p.d. across the capacitor and the resistor will always be the same. The charge Q on the capacitor will decrease exponentially with time. Since $Q = VC$ and $V = IR$, the p.d. across the capacitor and the current I in the circuit also decrease exponentially with time.

The charge Q, p.d. V and the current I at a time t can be calculated using the equations:

$$Q = Q_0 e^{-\frac{t}{CR}} \qquad V = V_0 e^{-\frac{t}{CR}} \qquad I = I_0 e^{-\frac{t}{CR}}$$

where Q_0, V_0 and I_0 are the charge, voltage and current at time $t = 0$ respectively. (e is the base of natural logs; equal to 2.718...)

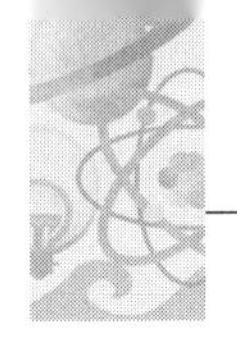

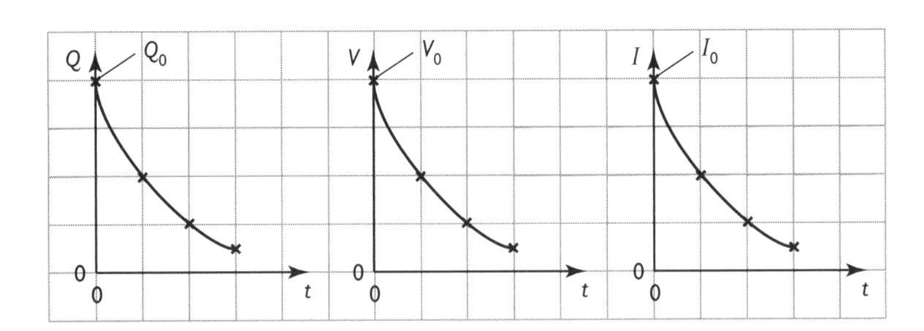

Note:

- The *gradient* of the charge Q against time t graph is equal to *current* $\left(I = \frac{\Delta Q}{\Delta t}\right)$.
- The *area* under the current I against time t graph is equal to *charge*.

Exponential decay

The charge stored by the capacitor decreases exponentially with time. This means that in a given time interval the *fraction* of the charge left on the capacitor is the same. For example, if in a 10 s period the charge stored drops to 80% of its initial value, then after the next 10 s the charge will again drop to 80% of its previous value, and so on. An exponential decay has a *constant-ratio property*.

Time constant

The product of capacitance and resistance, CR, appears in the equations for exponential decay of charge etc. What is the relevance of this product?

Consider what happens when the time t is equal to CR. Since $Q = Q_0 e^{-\frac{t}{CR}}$, we have:

$$Q = Q_0 e^{-1} \approx 0.368Q_0$$

The charge left on the capacitor decreases to e^{-1} or roughly 0.37 of its initial charge.

Since the charge decays exponentially, this means that in a time equal to CR the charge will always decrease to 0.37 of its previous value. This important idea leads to the concept of time constant of a circuit. The definition for time constant is as follows:

The time constant of a capacitor–resistor circuit is defined as the time taken for the charge (or the current or p.d.) to decrease to e^{-1} or approximately 0.37 of its previous value.

In exams, if you are asked to define time constant, then the definition is as stated above. However, for calculations it is helpful to know that time constant is also equal to CR. That is:

time constant = CR

Note: the product CR must have the same unit as time (seconds). This must also imply that 1 F Ω = 1 s.

Worked example

A capacitor of capacitance 120 μF is charged to 10 V. It is discharged through a resistor of resistance 100 kΩ. Calculate:

(a) the initial current in μA in the resistor

(b) the current in the circuit after 24 s

Answer

(a) The initial p.d. across the resistor is 10 V. Therefore:

$$I = \frac{V}{R} = \frac{10}{100 \times 10^3} = 1.0 \times 10^{-4}\,\text{A}$$

initial current = 100 μA

(b) time constant = CR

$= 120 \times 10^{-6} \times 100 \times 10^3 = 12\,\text{s}$

$$I = I_0 e^{-\frac{t}{CR}}$$

$$I = 100 \times e^{-\frac{24}{12}} \approx 13.5\ \mu\text{A}$$

There is an alternative method for (b). The time of 24 s is twice the time constant. The current will decrease to $(e^{-1})^2$ or $(0.368)^2$ of its initial value. Therefore:

current = $100/e^2$ = 13.5 μA

The nuclear atom

Atomic structure

An atom has a *positively* charged nucleus. The nucleus contains *neutrons* and *protons*. The neutron has no charge and the charge on a proton is positive and equal in magnitude to the elementary charge e. The nucleus is surrounded by a *cloud of electrons*. For a neutral atom, the number of protons is equal to the number of electrons.

The mass of a proton and a neutron is roughly the same. The electron is about 2000 times less massive than a proton. Hence, the majority of the mass of the atom is contained within the nucleus.

Here are some typical order of magnitude values for the size of particles:

- radius of the electron $< 10^{-18}$ m
- radius of the nucleus $\sim 10^{-14}$ m
- radius of the atom $\sim 10^{-10}$ m
- size of molecule $\sim 10^{-10}$ m to 10^{-7} m

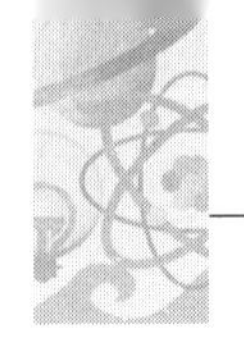

Alpha particle scattering experiment

Geiger and Marsden, working under the direction of Rutherford, carried out the alpha scattering experiment in 1911. This experiment led to the nuclear model of the atom.

The alpha particles (positive helium-4 nuclei) from a radioactive source were targeted towards a thin gold foil. The number of alpha particles scattered at various angles were counted using a scintillator screen mounted onto a microscope. The scintillator produced a tiny speck of light when it was struck by a high-speed alpha particle.

Observation	Conclusion
Most of the alpha particles were not scattered much	The gold atoms were mostly empty space (vacuum)
Some of the alpha particles were scattered through very large angles (>90°)	Each atom has a positive nucleus with a radius of about 10^{-14} m

Diagram (a) below shows a simplified arrangement of the experiment and diagram (b) shows the typical paths of alpha particles when coming very close to the gold nucleus.

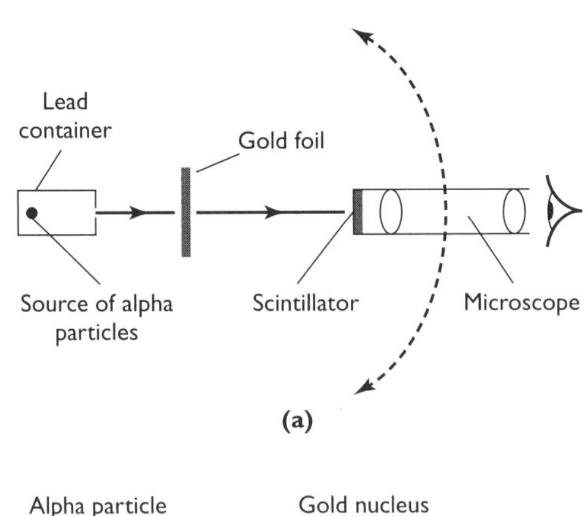

(a)

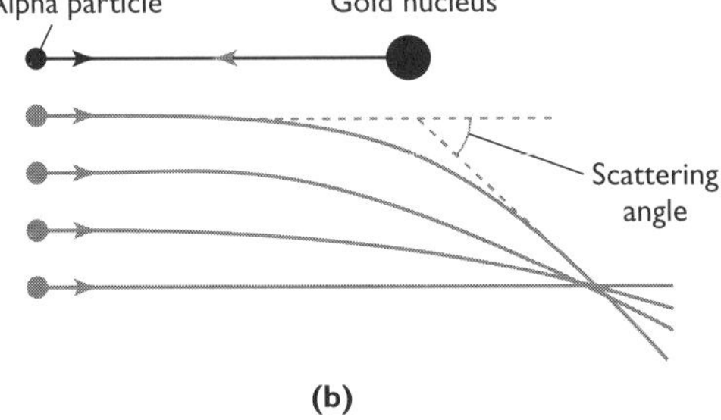

(b)

The force on the alpha particle can be determined using the equation for Coulomb's law. At a distance of about 10^{-14} m, the force on the alpha particle is about 360 N. Try this for yourself using $F = Qq/4\pi\varepsilon_0 r^2$, with $Q = 1.26 \times 10^{-17}$ C (the charge on gold nucleus), $q = 3.20 \times 10^{-19}$ C (the charge on an alpha particle), $\varepsilon_0 = 8.85 \times 10^{-12}$ F m^{-1} and $r = 10^{-14}$ m.

Nuclear density

The nucleus of an atom is very small and massive — hence the density of a nucleus is far greater than the density of matter, which is made up of atoms.

Consider copper, which has a density of 8900 kg m^{-3}. The *nucleus* of a copper atom has a radius about 5×10^{-15} m and mass 1.1×10^{-25} kg. The average nuclear density ρ_{nuc} of copper is:

$$\rho_{nuc} = \frac{\text{mass}}{\text{volume}}$$

$$= \frac{1.1 \times 10^{-25}}{\frac{4}{3}\pi \times (5 \times 10^{-15})^3} = 2.1 \times 10^{17}\ \text{kg m}^{-3} \sim 10^{17}\ \text{kg m}^{-3}$$

All nuclei have a density of about 10^{17} kg m^{-3}. This is roughly 10^{14} times greater than the density of ordinary matter (the density of water is 1000 kg m^{-3}). A spoonful of nuclear material has a mass of about 1000 million tonnes.

Strong nuclear force

The positively charged protons inside the nucleus of an atom strongly repel each other. What keeps them together inside the nucleus? The attractive gravitational force between the protons is too weak by a factor of about 10^{36}. The attractive force holding the protons together is the **strong nuclear force**. The neutrons within the nucleus also experience this force. The strong nuclear force:

- is an attractive force
- is a very short-range force (10^{-14} m)

The nucleus

Here are some helpful terms, definitions and ideas:

- The term **nucleon** means either a neutron or a proton.
- The **nucleon number** A is equal to the number of nucleons inside a particular nucleus — this is also sometimes known as the **mass number**.
- The **proton number** Z is equal to the number of protons inside a particular nucleus — this is also sometimes known as the **atomic number**.
- The charge on the nucleus = $+Ze$.
- A **nuclide** is a particular combination of neutrons and protons.
- **Isotopes** are nuclei of the same elements with a different number of neutrons but the same number of protons.

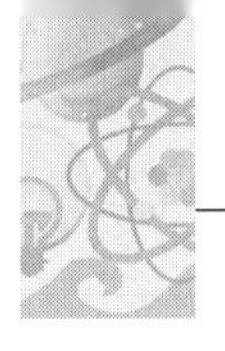

The nucleus of an atom is represented as $^{A}_{Z}X$, where X is the chemical symbol for the element. For example, an isotope of uranium with 92 protons and 146 neutrons is represented as $^{238}_{92}U$.

Fundamental particles

Hadrons and leptons

Fundamental particles cannot be subdivided into smaller constituents so they have no internal structure. **Quarks** and **electrons** are examples of fundamental particles. Particles are divided into two categories of hadrons and leptons.

- **Hadrons** are a group of particles that consist of **quarks**. All hadrons are affected by the **strong nuclear force**. The range of the strong nuclear force is about 10^{-14} m. Protons, neutrons and mesons are examples of hadrons.
- **Leptons** are a group of particles that are not affected by the strong nuclear force. Electrons, neutrinos and muons are examples of leptons.

Simple quark model

Quarks account for the properties (mass, charge, spin etc.) of the hadrons. In the simple quark model, there are six quarks and six corresponding antiquarks. The properties of the quarks are described by their charge, *Q*, baryon number, *B*, and strangeness, *S*.

Quark	Q (in units of e)*	B	S
Up (u)	$+\frac{2}{3}$	$+\frac{1}{3}$	0
Down (d)	$-\frac{1}{3}$	$+\frac{1}{3}$	0
Strange (s)	$-\frac{1}{3}$	$+\frac{1}{3}$	–1
Charm (c)	$+\frac{2}{3}$	$+\frac{1}{3}$	0
Top (t)	$+\frac{2}{3}$	$+\frac{1}{3}$	0
Bottom (b)	$-\frac{1}{3}$	$+\frac{1}{3}$	0

* Elementary charge $e = 1.6 \times 10^{-19}$ C

Hadrons revisited

An antiquark has a 'bar' over the top of the symbol. For example, a down antiquark is represented as $\bar{d}$.

A **proton** has two up quarks and a down quark.

Hence, proton = (u u d).

Use the table on p. 40 to convince yourself that $Q = 1$, $B = 1$ and $S = 0$.

A **neutron** has one up quark and two down quarks.

Hence, neutron = (u d d).

This combination of quarks gives $Q = 0$, $B = 1$ and $S = 0$.

A **pi⁺ meson** has one up quark and one down antiquark.

Hence, pi^+ meson = (u $\bar{d}$).

This combination of quarks gives $Q = 1$, $B = 0$ and $S = 0$.

In all nuclear reactions, charge, Q, baryon number, B, and strangeness, S, are conserved. You can use the table to show that this is true for the following nuclear reaction:

$$\underset{(u\,u\,d)}{\text{proton}} + \underset{(u\,u\,d)}{\text{proton}} \rightarrow \underset{(u\,u\,d)}{\text{proton}} + \underset{(u\,d\,d)}{\text{neutron}} + \underset{(u\,\bar{d})}{\text{pi}^+ \text{ meson}}$$

Radioactivity

Natural radioactivity is the *spontaneous* and *random* disintegration of unstable nuclei because they have surplus energy. In the process of disintegration, a nucleus emits a particle (alpha particle, α, beta-minus, β^-, or beta-plus, β^+) and/or a gamma ray photon (γ).

Radioactive decay is a *spontaneous* event. This means that the decay of nuclei is unaffected by:

- external factors such as pressure, temperature or even chemical reactions
- the presence of other nuclei

Radioactivity is also a *random* event. This means that:

- it is not possible to predict when a particular nucleus in a sample will decay
- each nucleus of a particular isotope has the same probability of decay per unit time

Basic properties and characteristics of radiations

Alpha particles

An alpha particle is a nucleus of helium with a charge of $+2e$.

Alpha particles:

- are emitted with roughly the same kinetic energy, typically 1 MeV
- have a speed of about $10^6\,\text{m}\,\text{s}^{-1}$
- are strongly ionising
- have a short range of about a few centimetres in air
- are deflected by both electric and magnetic fields

Some **smoke detectors** use a weak alpha source (americium-241) to detect the presence of smoke particles.

Beta-minus particles

A beta-minus particle is an electron with a charge of $-e$.

Beta particles:

- are emitted with a range of kinetic energies (because of the energy and momentum of the antineutrino)
- have a speed about $10^7\,\text{m}\,\text{s}^{-1}$
- are less ionising than alpha particles
- have a range of a few millimetres in aluminium
- are deflected by both electric and magnetic fields

Gamma rays

Gamma rays are short-wavelength electromagnetic waves.

Gamma rays:

- have no charge
- travel at the speed of light ($3.0 \times 10^8\,\text{m}\,\text{s}^{-1}$)
- are weakly ionising
- have a range of a few centimetres in lead
- are not affected by either electric or magnetic fields

Stability of nuclei

The stability of a nucleus is linked to the relative number of neutrons, N, and the number of protons, Z. The diagram below shows a graph of N against Z.

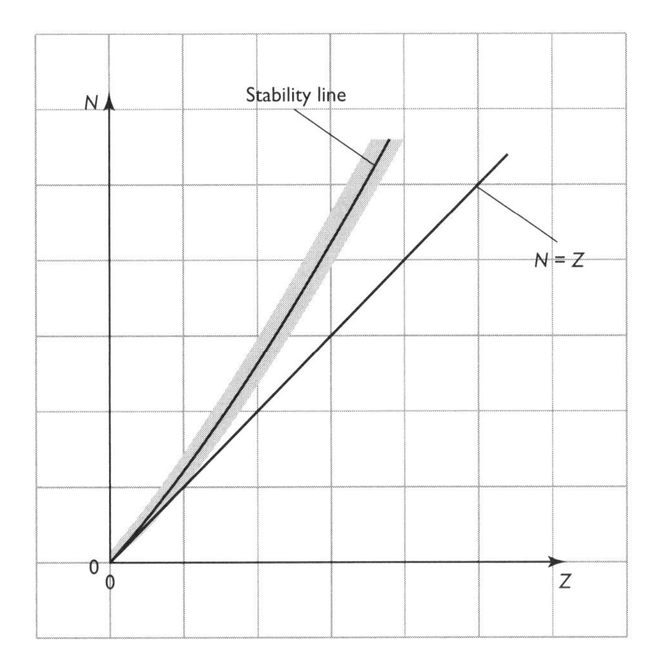

Stable nuclei lie on the line of stability. The shaded regions on either side of this line show the location of unstable nuclei. The decay of an unstable nucleus brings it closer to the line of stability.

Alpha decay

In an **alpha decay**, a helium nucleus (two protons and two neutrons; ${}^{4}_{2}\text{He}$) is emitted from the unstable nucleus. In this process, the nucleon number decreases by 4 and the proton number decreases by 2. For example:

$${}^{232}_{90}\text{Th} \rightarrow {}^{228}_{88}\text{Ra} + {}^{4}_{2}\text{He}$$

Note: A and Z numbers are conserved.

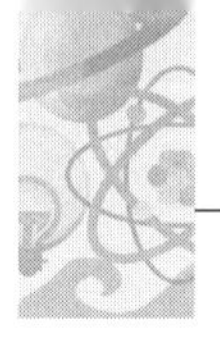

Beta-minus decay

Some neutron-rich nuclei emit an electron ($^{0}_{-1}e$) in a decay known as **beta-minus (β^-) decay**. In this process, the nucleon number remains unchanged and the proton number increases by 1. For example:

$$^{90}_{38}Sr \rightarrow {}^{90}_{39}Y + {}^{0}_{-1}e \quad (+ \bar{\nu})$$

Note: *A* and *Z* numbers are conserved.

The *weak nuclear force* is responsible for this decay. A neutron ($^{1}_{0}n$) within the nucleus transforms into a proton ($^{1}_{1}p$), an electron ($^{0}_{-1}e$) and an antineutrino ($\bar{\nu}$). That is:

$$^{1}_{0}n \rightarrow {}^{1}_{1}p + {}^{0}_{-1}e + \bar{\nu}$$

The antineutrino has no charge and very little mass.

At a quark level, a down quark transforms into an up quark, an electron and an antineutrino.

$$d \rightarrow u + {}^{0}_{-1}e + \bar{\nu}$$

Beta-plus decay

Some proton-rich nuclei emit a positron ($^{0}_{+1}e$) in a decay known as **beta-plus (β^+) decay**. In this process, the nucleon number remains unchanged and the proton number decreases by 1. For example:

$$^{15}_{8}O \rightarrow {}^{15}_{7}N + {}^{0}_{+1}e \quad (+ \nu)$$

Note: *A* and *Z* numbers are conserved.

The *weak nuclear force* is also responsible for this decay. A proton ($^{1}_{1}p$) within the nucleus transforms into a neutron ($^{1}_{0}n$), a positron ($^{0}_{+1}e$) and an neutrino (ν). That is:

$$^{1}_{1}p \rightarrow {}^{1}_{0}n + {}^{0}_{+1}e + \nu$$

At a quark level, an up quark transforms into a down quark, a positron and a neutrino.

$$u \rightarrow d + {}^{0}_{+1}e + \nu$$

Rules of radioactive decay

Here are two important definitions for activity and decay constant.

The activity of sample is the rate at which the nuclei decay or disintegrate.

The decay constant is the probability of decay of a nucleus per unit time.

The SI unit of activity is the becquerel (Bq). 1 Bq = 1 disintegration per second.

Candidates often misunderstand the activity of a sample. For example, what is meant by a radioactive sample having an activity of 1000 Bq? There are two interpretations.

You can imagine 1000 nuclei decaying per second or 1000 α particles or β particles or γ-ray photons being emitted per second.

Consider a radioactive sample with an activity of 2.0×10^{10} Bq, emitting α particles each of kinetic energy 1.5×10^{-13} J. We can determine the power of the sample as follows:

power = rate of energy emitted from sample

= number of α particles emitted per second $\times$ energy of each α particle

= activity $\times$ 1.5×10^{-13} J

= $2.0 \times 10^{10} \times 1.5 \times 10^{-13}$

= 3.0×10^{-3} W

The activity A of a sample of radioactive material is directly proportional to the number N of undecayed nuclei and is related to the decay constant λ by the following equation:

$$\boldsymbol{A = \lambda N}$$

The unit of decay constant is normally s^{-1}. It would be incorrect to write this unit as becquerel, Bq, or hertz, Hz. The unit of decay constant can also be min^{-1}, h^{-1} or y^{-1}.

Decay equations

The activity A of a sample decays *exponentially* with time. The count rate C recorded by a counter connected to a Geiger–Müller tube placed close to the radioactive sample will be a fraction of the total activity. Consequently, the count rate will decay exponentially. Since activity is directly proportional to the number N of active or undecayed nuclei in the sample, N must also decay exponentially.

The activity, A, and number, N, of undecayed nuclei and the count rate, C, at a time t can be calculated using the equations:

$$\boldsymbol{A = A_0 e^{-\lambda t}} \qquad \boldsymbol{N = N_0 e^{-\lambda t}} \qquad \boldsymbol{C = C_0 e^{-\lambda t}}$$

where A_0, N_0 and C_0 are the activity, number of undecayed nuclei and count rate at time $t = 0$ respectively. e is the base of natural logs, equal to 2.718...

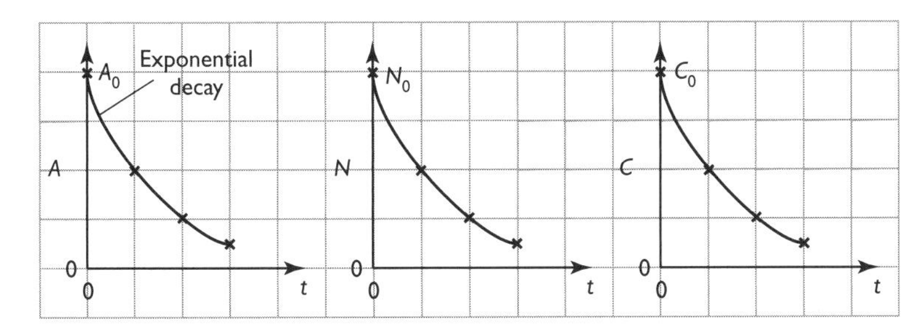

Note: these equations are similar to the exponential decay of charge for a capacitor–resistor circuit. Hence, each graph will show a *constant-ratio property* (see p. 36).

Half-life

The rate of decay of a radioactive sample depends on the half-life of its isotope.

The half-life $t_{1/2}$ of an isotope is the average time taken for half of the active nuclei to decay.

Even isotopes of the same element can have an extraordinary range of half-lives. This is illustrated in the table below for carbon.

Isotope of carbon	Half-life
Carbon-10	19 s
Carbon-11	20.5 min
Carbon-12	Infinite (stable)
Carbon-13	Infinite (stable)
Carbon-14	5570 years
Carbon-15	2.3 s

The isotope carbon-14 is used in **carbon dating** to determine the age of relics that contain carbon, such as paper, clothing, bones etc.

The half-life, $t_{1/2}$, of an isotope is related to the decay constant, λ, by the equation:

$$\lambda t_{1/2} = \ln(2) \approx 0.693$$

The graph below shows the variation of the number, N, of active nuclei left in a sample with time t.

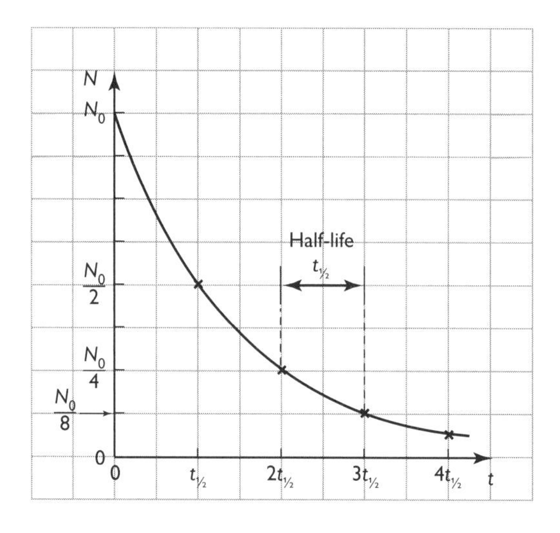

Exponential decay of the number of active nuclei

Note:

- The graph has a constant-ratio property; the number of active nuclei is halved after a time interval equal to one half-life.
- The *fraction* of the active nuclei remaining after *n* half-lives is equal to $\left(\frac{1}{2}\right)^n$.

Worked example

The isotope thorium-234 has a half-life of 6.7 hours. The initial activity of a thorium sample is 8.0×10^9 Bq. Calculate:

(a) the initial number of undecayed nuclei of thorium in the sample

(b) the activity of the sample after 24.0 hours

Answer

(a) In order to find the initial number, N_0, of active nuclei, we need to calculate the decay constant, λ.

$\lambda t_{1/2} = \ln(2) \approx 0.693$

$$\lambda = \frac{0.693}{(6.7 \times 3600)} = 2.87 \times 10^{-5}\,\text{s}^{-1}$$

$A = \lambda N$

Therefore:

$$N_0 = \frac{A_0}{\lambda} = \frac{8.0 \times 10^9}{2.87 \times 10^{-5}} = 2.78 \times 10^{14}$$

(b) The activity of the sample decays exponentially; $A = A_0 e^{-\lambda t}$. Therefore:

$$A = 8.0 \times 10^9 \times e^{-(2.87 \times 10^{-5} \times 24 \times 3600)}$$

$$A = 8.0 \times 10^9 \times 0.0838 \approx 6.7 \times 10^8\ \text{Bq}$$

In part (b) it is vital that you have the decay constant in s^{-1} because the activity is given in Bq.

Nuclear fission and fusion

Einstein's mass–energy equation

The mass–energy equation is:

$$\Delta E = \Delta m c^2$$

where ΔE is the change in energy of a system, Δm is the change in mass of the system and c is the speed of light in a vacuum. According to this famous equation, mass and energy are equivalent. In nuclear reactions, it is 'mass–energy' that is conserved and not just 'energy'.

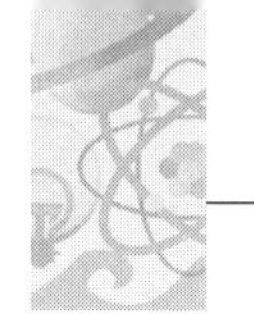

Here are two consequences of the mass–energy equation:

- The mass of a system *increases* when external energy is *supplied* to the system.
- Energy is *released* from the system when its mass *decreases*.

The 'system' could be a decaying radioactive nuclei, an accelerated electron, a person etc. For a person, the change in mass for everyday events is too small to be detected. For example, the increase in the mass of a 70 kg person travelling in a jet plane at 600 m s^{-1} is:

$$\Delta m = \frac{\Delta E}{c^2} = \frac{\frac{1}{2} \times 70 \times 600^2}{(3.0 \times 10^8)^2} = 1.4 \times 10^{-10} \text{ kg} \quad (\Delta E = \frac{1}{2}mv^2)$$

This miniscule change in mass is not going to be noticed by the person. However, the change in mass of a high-speed particle is much more significant. An electron travelling at 20% of the speed of light will have a 2% increase in its mass.

Revisiting natural radioactive decay

In natural radioactive decay, an unstable nucleus spontaneously emits a beta or alpha particle and/or a photon. We can use the mass–energy equation to understand how a nucleus emits energy. This is illustrated below.

A nucleus of carbon-14 decays by beta-minus emission according to the following decay equation:

$$^{14}_{6}\text{C} \rightarrow {}^{14}_{7}\text{N} + {}^{0}_{-1}\text{e} + \overline{\text{v}}$$

The carbon-14 nucleus can be assumed to be at rest at the start. In this decay, the nitrogen-14 nucleus recoils, and the beta particle and the neutrino fly off in the opposite direction. These particles have kinetic energy. Where does this kinetic energy comes from? In this decay, there is a *decrease* in mass of the system. According to the mass–energy equation, this implies that energy is *released* in the decay as kinetic energy.

Particle	**Mass/kg**
$^{14}_{6}$C nucleus	$2.3253914 \times 10^{-26}$
$^{14}_{7}$N nucleus	$2.3252723 \times 10^{-26}$
$^{0}_{-1}$e (electron)	$0.0000912 \times 10^{-26}$
$\overline{\text{v}}$ (anti-neutrino)	Negligible

initial mass of system = $2.3253914 \times 10^{-26}$ kg

final mass of system = $(2.3252723 \times 10^{-26} + 0.0000912 \times 10^{-26}) = 2.3253635 \times 10^{-26}$ kg

change in mass, $\Delta m = (2.3253635 \times 10^{-26} - 2.3253914 \times 10^{-26}) = -2.79 \times 10^{-31}$ kg

The minus sign simply shows that the mass decreases in this decay.

The kinetic energy ΔE of the particles after the decay is given by:

$$\Delta E = \Delta mc^2 = 2.79 \times 10^{-31} \times (3.0 \times 10^8)^2 \approx 2.5 \times 10^{-14}\,\text{J}$$

The kinetic energy of 2.5×10^{-14} J is equivalent to about 160 keV.

Nuclear binding energy

For all stable or unstable nuclei, it is found that the mass of each nucleus is always less than the total mass of its separate nucleons (protons and neutrons). That is:

mass of nucleus < total mass of the protons and neutrons

The mass of the carbon-12 nucleus is $1.9926483 \times 10^{-26}$ kg and the total mass of its 6 protons and 6 neutrons is $2.0085312 \times 10^{-26}$ kg. The difference is mass is known as **mass defect** and this is linked to the **binding energy** of the nucleus.

The mass defect of a nucleus is the difference in mass of all its separate protons and neutrons and the mass of the nucleus.

The binding energy of a nucleus is the minimum energy needed to separate all its nucleons.

It is not surprising that we can use Einstein's mass–energy equation to calculate the binding energy of the nucleus. In order to compare the stability of the different nuclei, we often need to consider the *binding energy per nucleon*.

The calculation below shows how we can do this for carbon-12 nucleus.

$$\text{mass defect for carbon-12 nucleus} = \text{mass of nucleons} - \text{mass of nucleus}$$

$$= (2.0085312 \times 10^{-26} - 1.9926483 \times 10^{-26})\,\text{kg}$$

$$= 1.58829 \times 10^{-28}\,\text{kg}$$

$$\text{binding energy} = \Delta mc^2 = 1.58829 \times 10^{-28} \times (3.0 \times 10^8)^2 \approx 1.43 \times 10^{-11}\,\text{J}$$

$$\text{binding energy per nucleon} = \text{binding energy/number of nucleons}$$

$$= \frac{1.43 \times 10^{-11}}{12} \approx 1.2 \times 10^{-12}\,\text{J}$$

The binding energy per nucleon for carbon-12 nucleus is about 1.2×10^{-12} J or about 7.5 MeV per nucleon. This means that on average, a minimum of 7.5 MeV is needed by each nucleon to free itself from the strong nuclear attractive forces of the rest of the nucleons.

A nucleus with a larger value of binding energy than another nucleus is much more stable. The binding energy per nucleon (BE/nucleon) against nucleon number A is shown in the graph below.

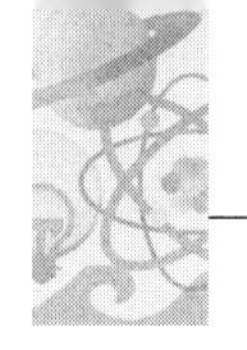

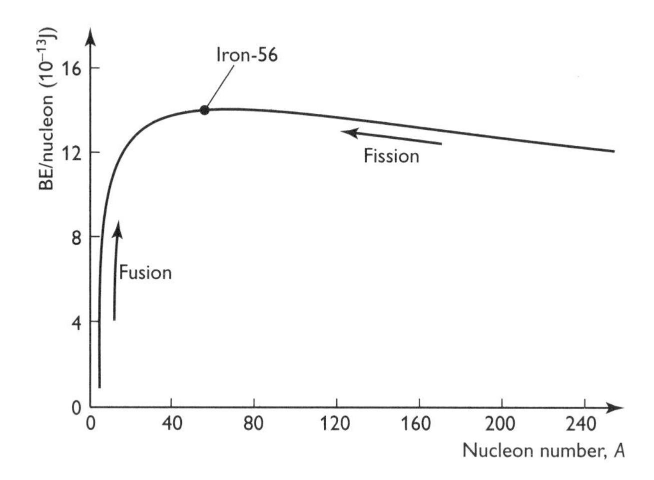

Note:

- Iron-56 is the most stable isotope in nature. It has the largest binding energy per nucleon.
- In all natural radioactive decays, the parent nucleus decays into a much more stable daughter nucleus.

Nuclear fission

A nucleus of uranium-235 can undergo an induced fission reaction. A fission reaction can be described as follows:

- A *slow-moving* neutron is absorbed by a nucleus such as uranium-235.
- The resulting nucleus is very unstable and splits into two unequal nuclei and number of *fast-moving* neutrons.

A typical fission reaction of uranium-235 is shown below:

$$ {}^{1}_{0}n + {}^{235}_{92}U \rightarrow \underbrace{{}^{236}_{92}U^{*}}_{\text{highly unstable}} \rightarrow {}^{90}_{36}Kr + {}^{143}_{56}Ba + 3\,{}^{1}_{0}n $$

In this fission reaction:

- The proton number, nucleon number and mass–energy are all conserved.
- The total mass of ${}^{90}_{36}Kr$, ${}^{143}_{56}Ba$ and the three neutrons is *less* than the total mass of ${}^{1}_{0}n$ and ${}^{235}_{92}U$.
- The energy released as kinetic energy of the neutrons and the *fragment* nuclei can be calculated from the decrease in mass Δm by using Einstein's mass–energy equation $\Delta E = \Delta mc^2$.

In most fission reactions, there are normally two, three or four neutrons emitted. These can go onto to create further fission reactions causing a **chain reaction**, as illustration below.

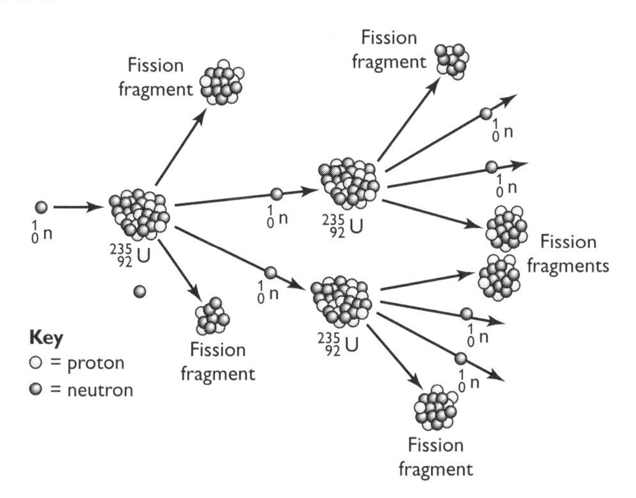

In a nuclear bomb, the chain reaction grows uncontrollably because of the exponential growth in the number of neutrons.

In a nuclear power station, the reactions are controlled. On average, a single neutron is absorbed by a uranium-235 nucleus between subsequent fission reactions.

Fission reactors

The diagram below shows the key components of a nuclear fission reactor.

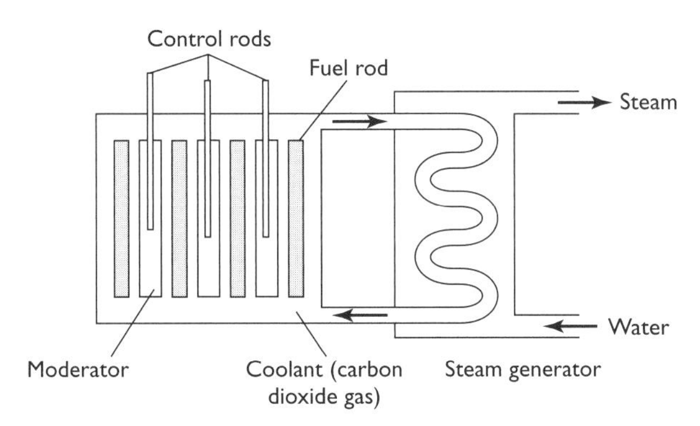

- **Fuel rods**: these contain pellets of *fissile material* (uranium, plutonium etc).
- **Coolant**: the coolant can be either gas (e.g. carbon dioxide) or liquid (e.g. water). The coolant is used to remove the thermal energy produced from the fission reactions in the reactor core. The thermal energy is used to heat water and create high-pressure steam to drive the turbines of the generators.

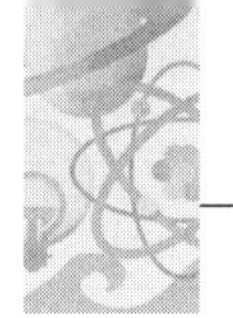

- **Moderator**: the nuclear fuel rods are surrounded by the moderator. A common material for the moderator is graphite. The purpose of the moderator is to slow down the fast-moving neutrons produced in the fission reactions. The fast-moving neutrons make inelastic collisions with the graphite atoms. Fast-moving neutrons have a smaller chance of reacting with the uranium nuclei than slow-moving ones.
- **Control rods**: the purpose of the control rods is to absorb the neutrons. The control rods can be lowered into the core to slow down the fission reactions. The control rods contain either boron or cadmium.

Nuclear waste from reactors

Nuclear waste from nuclear reactors cannot be disposed off as normal waste because it remains highly radioactive for an extremely long period. Radioactivity cannot be destroyed by burning the material. Burial of low-level waste in old mine-shafts is one method of containment. This may have a severe environmental impact if the waste leaks out, polluting underground water. Plutonium is one of the most dangerous waste products from nuclear reactors. It has a half-life of 24 000 years and is extremely toxic.

Nuclear fusion

Massive nuclei such as uranium can become much more stable by either fission reactions or natural radioactive decay. In a similar way, light nuclei such as deuterium ($^2_1$H) can become much more stable by joining together in a process known as **fusion**.

A typical fusion reaction between two deuterium nuclei is shown below:

$$^2_1\text{H} + {}^2_1\text{H} \rightarrow {}^4_2\text{He}$$

In this fusion reaction:

- The proton number, nucleon number and mass–energy are all conserved.
- The mass of the $^4_2$He nucleus is less than the total mass of the two $^2_1$H nuclei.
- The energy released as kinetic energy of the $^4_2$He nucleus can be calculated from the decrease in mass Δm by using Einstein's mass–energy equation $\Delta E = \Delta mc^2$.

Fusion reactions are not as easy to start compared with fission reactions. The uncharged neutron easily finds its way to a uranium nucleus because there is no electrostatic repulsion. The electrically charged deuterium nuclei repel each other, so they do not easily get close enough for the strong nuclear force to fuse them together. One way of achieving fusion is to heat the deuterium and increase the kinetic energy of the nuclei. At temperatures around 10^7 K, the deuterium nuclei have sufficient kinetic energy to overcome the electrostatic repulsion.

Stellar fusion

Nuclear fusion reactions occur in the cores of the stars, where hydrogen nuclei fuse together to produce helium nuclei. Thermonuclear fusion reactions are also responsible for the creation of elements such as lithium, carbon, oxygen and iron. The oxygen we breathe and the iron used to make our cars were created in stars that exploded a long time ago as supernovae.

Thermonuclear reactions take place in the core of stars. The chance of fusion reactions is greater because of:

- high temperatures (~10^7 K) — the hydrogen nuclei have greater kinetic energy
- high density (~10^5 kg m^{-3}) — the hydrogen nuclei are squashed together

X-rays

Producing X-rays

X-rays are electromagnetic waves with wavelengths in the range 10^{-8} m to 10^{-13} m. X-rays are produced when fast-moving electrons smash into a 'target' metal such as tungsten. The diagram below shows an X-ray tube.

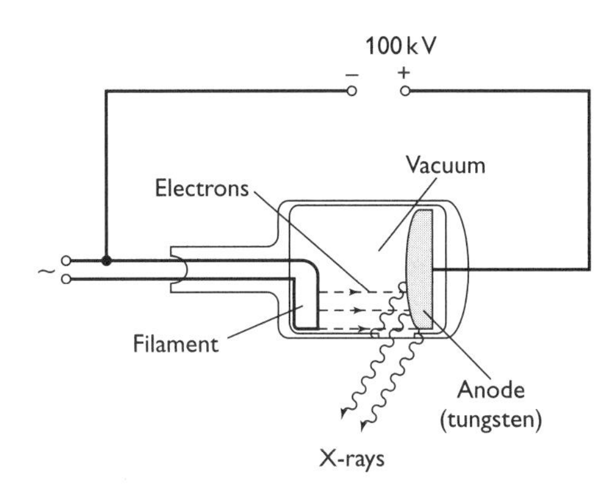

The electrons are produced at the hot filament. These electrons are accelerated towards the target metal, which is a high-melting point material such as tungsten. The potential difference between the cathode (filament) and anode (target metal) is typically 100 kV. Most of the kinetic energy of the electrons is transformed into heat in the metal. The target metal gets very hot and has to be cooled down by circulating water through it. About 1% of the kinetic energy of the electrons is converted into X-rays as photons.

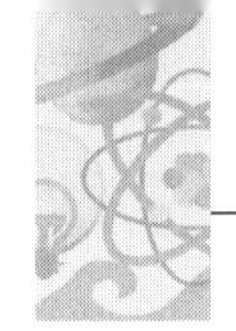

Worked example

Calculate the minimum wavelength of X-rays produced when an X-ray tube is operated at 200 kV.

Answer

The energy of a single electron is transformed into a single photon of X-rays. The maximum kinetic energy of the electron is Ve, where V is the p.d. and e is the elementary charge. The energy of a photon is given by the equation $E = \frac{hc}{\lambda}$. Therefore:

$$Ve = \frac{hc}{\lambda}$$

$$\lambda = \frac{hc}{Ve} = \frac{6.63 \times 10^{-34} \times 3.0 \times 10^{8}}{200 \times 10^{3} \times 1.60 \times 10^{-19}}$$

$$\lambda \approx 6.2 \times 10^{-12}\ \text{m}$$

This is a typical 'synoptic' question because it requires knowledge of topics from Unit G482.

Interaction mechanisms of X-rays

The *intensity* of an X-ray beam drops when passing through matter (e.g. bone and tissue). This is equivalent to X-ray photons either being stopped or scattered by the atoms of the material.

The table below summarises the three interaction (or attenuation) mechanisms by which X-ray photons are absorbed or scattered, as they pass through matter.

Interaction mechanism	Energy of X-ray photons/MeV	What happens to the X-ray photon?
Photoelectric effect	< 0.1	The photon disappears and its energy is used to eject an electron from an atom of the target metal
Pair production	0.5–5.0	The incident photon loses some of its energy to eject an electron from the atom of the target metal – the scattered photon has less energy (longer wavelength)
Compton effect	> 1.02	The photon disappears and produces an electron–positron pair

Intensity of X-rays

The intensity of an X-ray beam is defined as follows:

Intensity is the power per unit cross-sectional area.

We can determine the intensity I of an X-ray beam using the equation:

$$I = \frac{P}{A}$$

where P is the power and A is the cross-sectional area. The unit of intensity is watts per square metre, $W\,m^{-2}$.

The intensity of an X-ray beam decreases because of the interaction of the X-ray photons with the atoms of the material. The intensity I of an X-ray beam decreases exponentially with the thickness x of the material and is given by the equation:

$$I = I_0 e^{-\mu x}$$

where I_0 is the initial intensity, x is the thickness of the material and μ is the attenuation (or absorption) coefficient. The unit for the attenuation coefficient is m^{-1} (or cm^{-1} etc.).

Bone has a larger attenuation coefficient than soft tissues. This is why we can clearly identify bones rather than soft-tissues on an X-ray image (see Worked example below).

Worked example

X-ray photons of energy 50 keV are used when imaging the human skeleton. The attenuation coefficients for bone and muscle are $3.3\,cm^{-1}$ and $0.50\,cm^{-1}$ respectively. Calculate the fraction of the X-ray intensity after passing through 2.5 cm of:

(a) bone

(b) muscle

Comment on your values.

Answer

The intensity is given by the equation $I = I_0 e^{-\mu x}$.

Hence the fraction of the intensity after a thickness x of material is given by:

$$\text{fraction} = \frac{I}{I_0} = e^{-\mu x}$$

(a) For bone, the fraction = $e^{-(3.3 \times 2.5)} = 2.6 \times 10^{-4}$ (0.026%)

(b) For muscle, fraction = $e^{-(0.50 \times 2.5)} = 0.29$ (29%)

The bones have a larger attenuating coefficient than muscle and hence absorb a greater amount of X-ray photons. This is why is easier to image broken bones.

n Note: since the maximum energy of the X-ray photons is 50 keV, the attenuation mechanism must be the photoelectric effect (see table on p. 54).

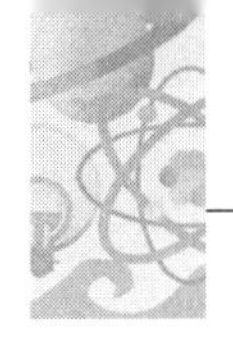

Detection systems

Intensifier screens

An intensifier screen may be used to reduce the exposure time for an X-ray image by a factor of 100 or more. The screen consists of a photographic plate/film sandwiched between two intensifier screens. Each screen is made from a material such as phosphor. Phosphor is a scintillator. The energy of a single X-ray photon incident on the scintillator is changed to several thousands of visible light photons, which produces a brighter image.

Image intensifiers

A diagram of an image intensifier is shown below.

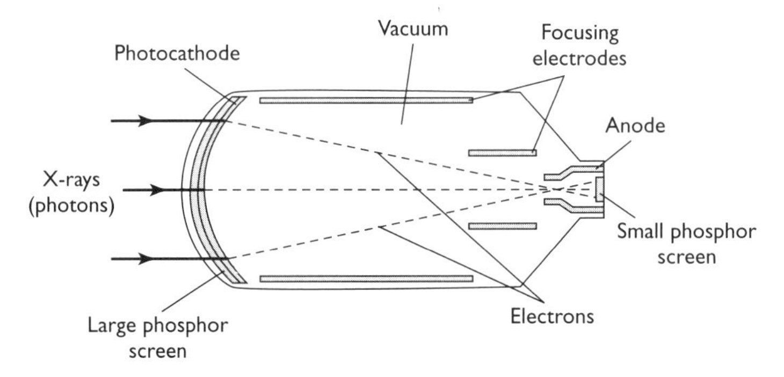

- The X-rays are incident on the first phosphor screen. The X-ray photons produce greater number of visible light photons.
- The visible light photons strike a photocathode where electrons are released.
- The electrons are accelerated and focused by the positively charged anode so they strike another phosphor screen, which then gives out visible light.
- The final image can be viewed on a monitor.

Image intensifiers are useful because:

- the final image is brighter
- exposure time to X-rays can be significantly reduced
- real-time images can be formed (a technique known as fluoroscopy)

Contrast media

Contrast media include materials such as iodine and barium. Either the patient swallows a liquid rich in barium (a 'barium meal') or the material is injected into the patient.

- A contrast material readily absorbs X-rays because it has a large attenuation coefficient or large proton number (Z).
- A contrast medium is used for imaging the outline or shape of soft tissues such as intestines.

Computerised axial tomography (CAT)

A CAT scanner uses X-rays to produce a three-dimensional image through a patient. Here are some important ideas about a CAT scanner.

- The patient lies on a table that moves in and out of a round opening known as the 'gantry'.
- The ring inside the gantry contains a single X-ray tube and about 720 X-ray detectors.
- The X-ray tube moves at high speed around the patient.
- The X-rays pass though the patients and reach the detectors on the opposite side of the X-ray tube.
- The detectors monitor the intensity of the X-rays absorbed.
- The computer connected to the detectors records thousands of images or 'slices' through the patients. (The Greek word for slice is *tomos*.)
- In a modern scanner, the X-ray tube can make three rotations every second and the computer can record 200 slices through the patient.
- A scan can take from 10 to 30 minutes.

Single X-ray images are much quicker than CAT scans. However, CAT scans have several advantages, as they:

- produce a three-dimensional image of the patient
- can distinguish between tissues of quite similar attenuation coefficients
- show the precise position and shape of tumours

Diagnostic methods in medicine

Tracers

Medical tracers are radioactive elements or compounds that are either ingested or injected into a patient to investigate metabolic functions or blood flow. Some examples of medical tracers are given below.

Iodine-131

Iodine-131 is a beta emitter with a half-life of about 8.1 days. The iodine is injected into the patient. Healthy kidneys will pass the iodine through to the bladder; if not, the amount of iodine builds up in the kidney. Geiger counters are used to monitor the concentration of this radioactive iodine. For healthy kidneys, the count rate will rise and then fall. However, if there is a blockage, the count rate will rise and then stay constant. Iodine-131 is also used to investigate the function of the thyroid gland.

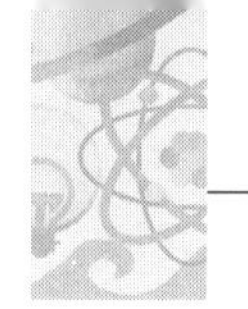

Technetium-99m

Technetium-99m is an extremely versatile tracer that can be used to monitor the function of the heart, liver, lungs, kidneys, brain etc. It has a short half-life of about 6 hours. Technetium-99m emits gamma rays. Gamma rays are the least ionising radiation and hence most of the gamma ray photons emitted by the tracer emerge out from the patient. The gamma ray photons are detected by a gamma camera, which pinpoints the location of the tracer within the patient, and hence help in diagnosing the function of the organ.

Gamma camera

A gamma camera is placed above the patient's organ. The diagram below shows the main components of the gamma camera.

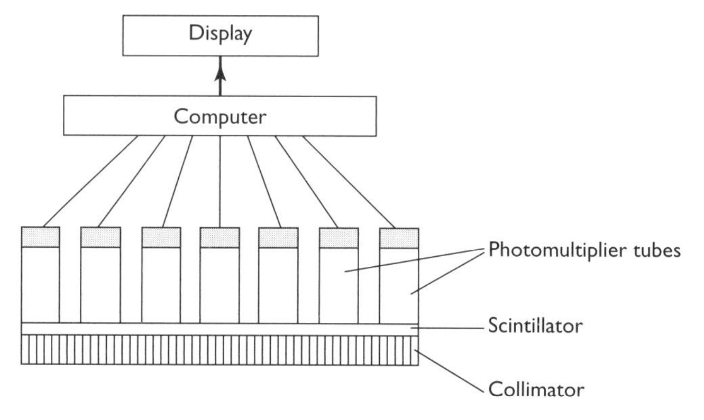

- **Collimator**: the collimator consists of a honeycomb of long, cylindrical lead tubes. Only gamma rays travelling along the axis of the tubes will reach the scintillator.
- **Scintillator**: each gamma ray photon incident at the scintillator produces an abundance of photons of visible light.

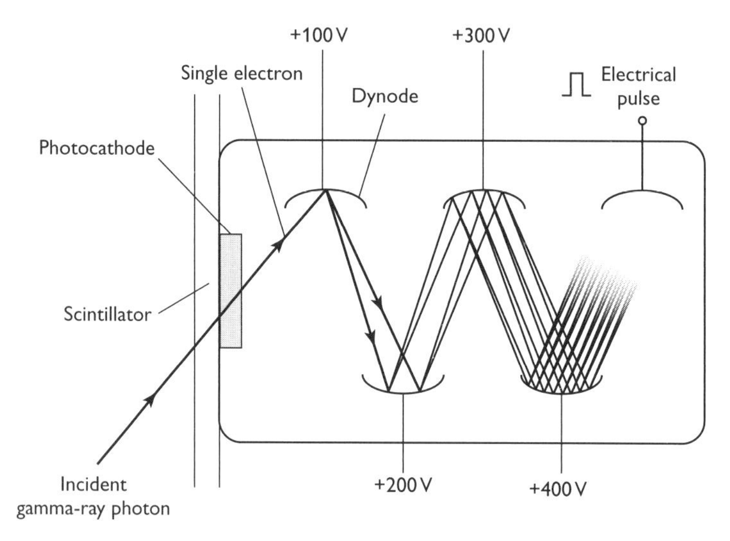

- **Photomultiplier tubes**: a single photon of visible light entering a photomultiplier tube produces a single electrical pulse. The photon of visible light produces a single electron from the photocathode by the photoelectric effect. This electron gains kinetic energy as it accelerates towards the first dynode. At this dynode the electron frees two or three secondary electrons. These in turn are accelerated towards other dynodes. The number of secondary electrons increases exponentially with the number of dynodes. For a photomultiplier tube with ten dynodes and an average of three secondary electrons produced at each dynode, there could be as many as $3^9 = 20\,000$ electrons at the final dynode. The electrons arriving at the final dynode represent an electrical pulse that can be registered by the computer.
- **Computer and display**: the electrical pulses from all the photomultiplier tubes are used to pinpoint the gamma-emitting tracer within the patient. Software is used to display the image on the monitor.

Positron emission tomography (PET)

PET is another technique that uses the fact that gamma rays emerge from a tracer within the patient. PET is similar to CAT because it too produces images of slices through the patient; one major difference is that it uses gamma rays and not X-rays.

The tracer used in PET is fluorine-18. ${}^{18}_{9}\text{F}$ is a beta-plus emitter, as shown below.

$${}^{18}_{9}\text{F} \rightarrow {}^{18}_{8}\text{O} + {}^{0}_{+1}\text{e} + \gamma + \nu$$

The positrons emitted in the decay of fluorine-18 are used in PET and not the gamma ray photons at this stage. Once emitted, the positron soon interacts with an electron and the two annihilate each other. The result of the annihilation is two gamma ray photons emitted in *opposite* directions.

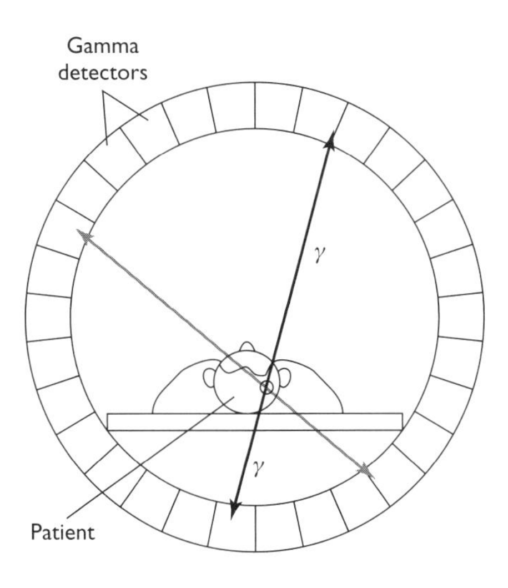

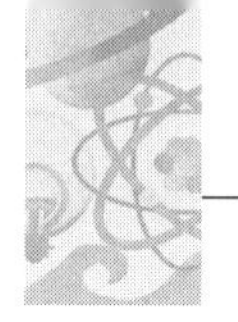

PET scans are used to monitor and diagnose the function of the brain. The patient is injected with a compound tagged by fluorine-19. In a PET scanner, the patient is placed in a ring consisting of gamma detectors. These detect *pairs* of gamma ray photons coming from inside the patient and travelling in *opposite* directions. The detectors are all connected to a computer. The arrival times of such photons are compared and computer is capable of pinpointing the exact location of the annihilation from the difference in the arrival times. Gradually, a three-dimensional image of the distribution of the tracer is built up and the abnormal function of the brain diagnosed.

Magnetic resonance

Many nuclei *spin* about an axis. This spin makes a nucleus (e.g. hydrogen or proton, carbon, phosphorus) behave as tiny magnets with north and south poles.

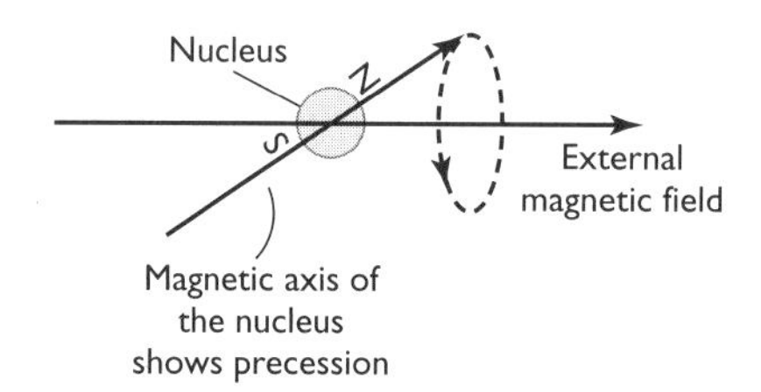

In the presence of a strong external magnetic field, two things happen:

- *Most* of the nuclei align their magnetic axes parallel to the external magnetic field (low-energy state) and a few nuclei align their magnetic axes anti-parallel to the external magnetic field (high-energy state).
- The magnetic axes of the nuclei rotate about the direction of the external field (just as a spinning top in the Earth's gravitational field) — they show **precession**.

The angular frequency of precession is known as the **Larmor frequency** ω_0. The precession frequency f_0 is directly proportional to the magnetic flux density B of the external magnetic field and is given by the equation:

$$f_0 = \frac{\gamma B}{2\pi}$$

where γ is the gyromagnetic ratio that depends on the nucleus itself. About 10% of our body mass is hydrogen, mostly contained in water and some in fats and proteins. The precession frequency for protons is about 60 MHz. It so happens that this frequency lies in the radio-wave region of the electromagnetic spectrum. Why is this so crucial?

Consider a patient lying in a strong magnetic field. Most of the protons within the patient precess about this field. When the patient is also subjected to a radio-wave signal of frequency equal to f_0, the low-energy state protons **resonate** and flip into a higher-energy state by precessing anti-parallel to the external magnetic field. When the radio-wave signal is switched off, the protons slowly return to their low-energy state and release their surplus energy as radio-wave photons of the same frequency f_0. The mean time taken for the protons to return from their high-energy state to the low-energy state is known as the **relaxation time**.

The relaxation time depends on the type of surrounding tissues. The relaxation time for watery tissues is several seconds and for fatty tissues about several hundred milliseconds. Cancerous tissues have intermediate times.

Magnetic resonance imaging (MRI)

The diagram below shows the main components of the magnetic resonance imaging (MRI) scanner.

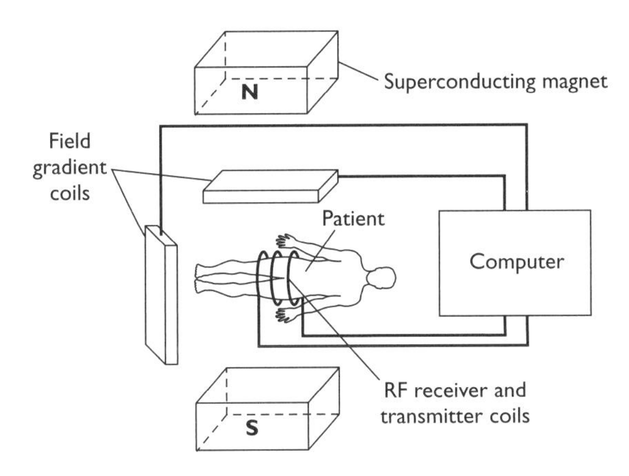

- **Superconducting magnet**: this provides the strong magnetic field of flux density about 2 T.
- **RF transmitter coil**: this coil transmits radio-frequency (RF) pulses into the patient.
- **RF receiver coil**: this coil picks up the radio waves emitted by the nuclei returning to their low-energy state.
- **Gradient coils**: these produce an additional external magnetic field along the length, depth and width of the patient. The Larmor frequency is therefore slightly different for each part of the body.
- **Computer**: the computer controls the radio-frequency pulses from the RF transmitter coils and analyses all the signals from the RF receiving coils. The location of the tissues is pinpointed by the slightly different Larmor frequencies due to the gradient coils. The type of tissue is identified from the different relaxation times. The computer produces images ('slices') through the patient and gives a detailed three-dimensional image.

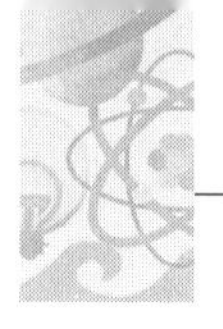

Advantages of MRI

- MRI scans produce better soft-tissue contrast than a CAT scan.
- An MRI scanner produces a three-dimensional image of the patient.
- The patient is not subject to ionising radiation, as with a CAT scan.
- Apart from the loud noise produced by the scanner during a scan, there are no after-effects.

Disadvantages of MRI

- Patient who have metallic objects such as surgical pins or pacemakers in their body cannot have an MRI scan as the metal objects become very hot.
- A single scan can take up to an hour; hence, the number of patients given MRI scans is far less than that for normal X-rays and CAT scans.

Ultrasound

What is ultrasound?

Ultrasound is any sound wave with a frequency greater than the upper limit of human hearing, i.e. above about 20 kHz. The typical frequency of ultrasound used for medical imaging is about 2 MHz. At such a high frequency, the wavelength of the ultrasound can be as small as 1 mm, which means smaller details can be detected in a scan. The speed of ultrasound depends on the material. The speed of ultrasound in water, muscle and bone is $1000\,\text{m}\,\text{s}^{-1}$, $1075\,\text{m}\,\text{s}^{-1}$ and $1600\,\text{m}\,\text{s}^{-1}$ respectively.

Ultrasound transducer

An ultrasound transducer is a device for emitting and detecting ultrasound. The key component of the transducer is the **piezoelectric crystal** (or film). It works on the principle of the *piezoelectric effect*. The same transducer is used to *transmit* and *detect* ultrasound.

- The crystal vibrates when an alternating voltage is applied between its ends. The frequency of vibration is the same as the frequency of the alternating voltage. The vibrations produce ultrasound from the vibrating crystal.
- The same crystal vibrates when ultrasound is incident on it. The forced vibration of the crystal induces an alternating e.m.f. across its end. The frequency of this induced e.m.f. is equal to that of the incident ultrasound.

Acoustic impedance

In an ultrasound scan, radiographers are interested in the intensity of ultrasound reflected at a boundary between two materials. This depends on the acoustic impedance of the materials. Acoustic impedance Z depends on the density ρ of the material and the speed c of ultrasound through the material.

Acoustic impedance is defined as follows:

acoustic impedance = density of material × speed of ultrasound in material

or

$Z = \rho c$

The unit of acoustic impedance is $kg\,m^{-2}\,s^{-1}$.

Reflected intensities

Consider an ultrasound incident at right angles to a boundary between two materials. The reflected intensity I_r is given by the equation:

$$\frac{I_r}{I_0} = \frac{(Z_2 - Z_1)^2}{(Z_2 + Z_1)^2}$$

or

$$\frac{I_r}{I_0} = \left(\frac{Z_2 - Z_1}{Z_2 + Z_1}\right)^2$$

where Z_1 and Z_2 are the acoustic impedances of the two materials.

Worked example

The acoustic impedance of bone and brain tissue is $6.4 \times 10^6\,kg\,m^{-2}\,s^{-1}$ and $1.6 \times 10^6\,kg\,m^{-2}\,s^{-1}$ respectively. With the aid of a calculation, explain why an ultrasound scan is not a sensible procedure for imaging a patient with suspected head/brain injuries.

Answer

For a bone–tissue boundary, the fraction of reflected intensity is:

$$\frac{I_r}{I_0} = \left(\frac{Z_2 - Z_1}{Z_2 + Z_1}\right)^2 \left(\frac{6.4 - 1.6}{6.4 + 1.6}\right)^2 = 0.36 \text{ (36\%)}$$

An ultrasound scan of the head will not give details of the brain because the skull will reflect a significant amount of the ultrasound intensity.

Note: the 10^6 does not have to be included in the calculation because it cancels out.

In an ultrasound scan, radiographers would like to distinguish between different tissues, but this is only possible if the acoustic impedances are very different.

Impedance (or acoustic) matching

The acoustic impedance of air is $400\,kg\,m^{-2}\,s^{-1}$ and that of skin is $1.7 \times 10^6\,kg\,m^{-2}\,s^{-1}$. When an ultrasound transducer is placed directly on to the skin of the patient, the value of ratio I_r/I_0 is about 0.9991. Hence, 99.91% of the incident ultrasound intensity is reflected back. This is not helpful if we want to image the internal structures of the

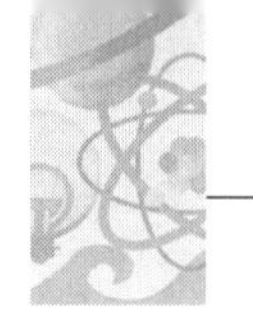

patient. In order to ensure most of the ultrasound is transmitted into the patient, a special gel is smeared between the patient's skin and the transducer. The acoustic impedance of the gel is almost the same as that of the skin — a process known as **impedance matching** or **acoustic matching**. Since $Z_2 \approx Z_1$, a very small amount of ultrasound is reflected back.

A-scans

A-scans are useful for determining the internal dimensions of cysts, organs, etc.

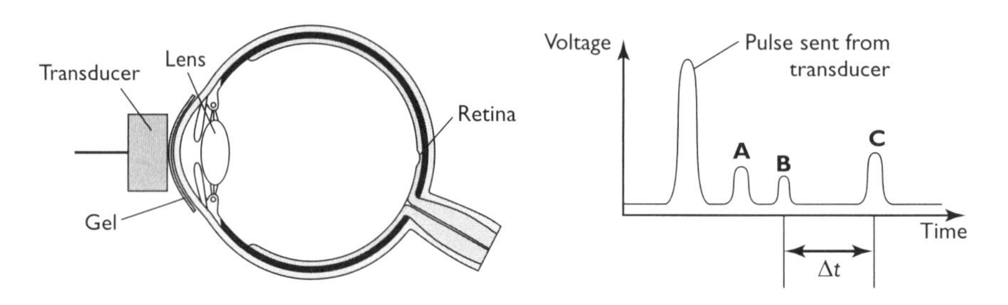

The ultrasound transducer sends pulses into the patient, and the same transducer detects the reflections off various boundaries. The simplified diagram above shows a scan of the eye. The reflected pulses A and B are from the front and back of the eye lens. Pulse C is reflected off the back of the eye (retina). The time Δt is equal to the time taken for ultrasound to travel *twice* the length between the retina and the back of the lens. The length between the retina and the lens can be determined from the speed of ultrasound within the fluid of the eye (see the worked example below).

Worked example

The speed of ultrasound in the fluid in the eye is $1060\,\text{m}\,\text{s}^{-1}$. A radiographer finds that the time taken for a pulse to travel twice the length of the eyeball is $40\,\mu\text{s}$. Use this information to determine the length of the eyeball.

Answer

total distance travelled by ultrasound $= v \times \Delta t$

$$= 1060 \times 40 \times 10^{-6} = 4.24 \times 10^{-2}\,\text{m}$$

This distance of 4.24 cm is twice the length of the eyeball. Hence:

$$\text{length of eyeball} = \frac{4.24}{2} \approx 2.1\,\text{cm}$$

B-scans

B-scans produce a detailed image of the cross-section through the patient, composed of many A-scans. The transducer is moved around the patient's body.

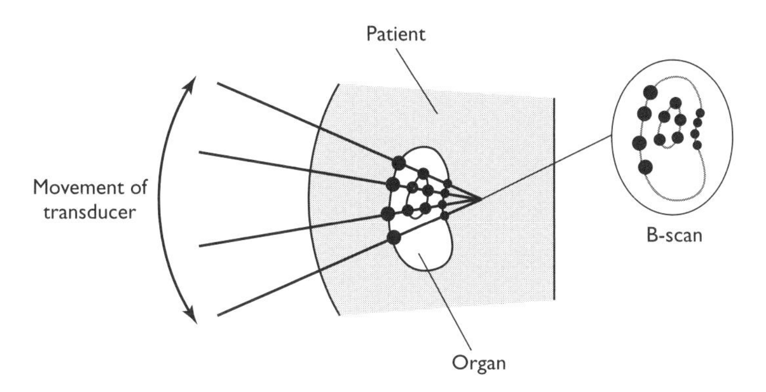

A computer works out the position and orientation of the transducer. Each reflected pulse is analysed and the *depth* and *nature* of the reflecting surfaces determined. A B-scan generates a two-dimensional image of the inside of the patient.

Application of the Doppler effect

The speed of blood in arteries and the rate of heartbeat can all be determined by using ultrasound. The phenomenon used is known as the Doppler effect.

In the Doppler effect, the wavelength or frequency of a wave changes when there is relative velocity between a source and detector.

Speed of blood flow

When measuring the speed of blood flow, an ultrasound transducer is pointed towards a major blood artery. The iron-rich blood cells reflect the ultrasound. For blood travelling towards the transducer, the reflected ultrasound has a slightly higher frequency. The change in the frequency Δf of the ultrasound is directly proportional to the speed of the blood. The speed of the blood can be determined from Δf.

Rate of heartbeats

An ultrasound transducer is pointed towards the heart. The surface of the heart approaches and recedes from the transducer. The increase and decrease in the frequency of the detected ultrasound can be used to determine the rate of the heartbeat.

Non-invasive techniques

In a non-invasive technique, the patient does not have to undergo surgery. Examples include:

- 2D X-ray imaging
- CAT scans
- tracers and gamma camera
- PET scans
- MRI scans
- ultrasound scans

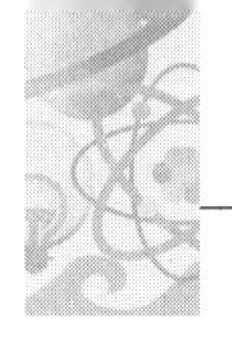

Structure of the universe

The solar system

Our **solar system** consists of the Sun, planets, asteroids, planetary satellites and comets. The Sun is a part of our galaxy known as the Milky Way.

The **universe** has about 10^{11} **galaxies**; each galaxy has about 10^{12} stars. It is also saturated by electromagnetic radiation (mainly in the microwave region of the spectrum), interstellar dust and dark matter. Most likely candidates for dark matter are neutrinos and black holes.

Stellar formation

All the stars burning in the night sky have had the same start in life. Their life began as a large interstellar gas cloud consisting mainly of atoms of hydrogen and small amounts of other elements including iron.

- Gravitational attraction between the atoms of the dust cloud causes the cloud to collapse.
- The **gravitational collapse** causes the gas cloud to heat up.
- The atoms in the dust cloud have greater kinetic energy and move faster.
- As the temperature of the cloud rises, the chance of **fusion reactions** between hydrogen increases.
- At a temperature around **10^7 K**, the hydrogen nuclei fuse together to produce helium nuclei. This reaction, known as **hydrogen burning** can be summarised as:

  $$4\,{}^{1}_{1}\text{H} \rightarrow {}^{4}_{2}\text{He} + 2\,{}^{0}_{+1}\text{e} + 2\nu$$

- Fusion reactions further increase the temperature of the cloud.
- A star of a stable size is formed when the gravitational pressure balances out the radiation pressure (from the photons released in the fusion reactions).
- The initial size of the star depends on the mass of the initial dust cloud. Our Sun is an average star of mass 2×10^{30} kg.

Stellar evolution

The final fate of a star depends on its mass. The core of an older star is layered with different shells or layers of elements. Iron is at the centre, followed by layers of silicon, oxygen, neon, carbon, helium and hydrogen.

When most of the fuel is used up by the star, the radiation pressure *decreases*. The increase in the gravitational pressure causes the helium nuclei in the outer layer to fuse together. The increase in the power production from the helium shell causes the outer layer of the star to *expand* due to radiation pressure.

For a star of mass < 3 solar masses

- The surface area of the star increases and its surface temperature drops. It becomes a **red giant**. (Even though the surface of the red giant is cooler, its output power is colossal because of its large surface area. Red giants appear as very bright red stars in the night sky.)
- The core of the star continues to collapse. When the temperature reaches about 10^8 K, the helium starts to fuse at a phenomenal rate. In a process known as **helium flash**, about half the material surrounding the core is ejected away as a **planetary nebula**.
- The remnant core left behind is known as a **white dwarf**. There are no further fusion reactions inside a white dwarf. It glows brightly because photons produced from past fusion reactions leak away from its hot surface. The density of a white dwarf is about 10^{15} kg m^{-3}.
- A white dwarf is prevented from gravitational collapse because of **electron degeneracy** or **Fermi pressure**. This comes about because two electrons cannot exist in the same quantum state. The *maximum* mass of a white dwarf is about 1.4 solar masses; this upper limit to the mass of a white dwarf is known as the **Chandrasekhar limit**.

For a star of mass > 3 solar masses

- The surface area of the star increases and its surface temperature drops. It becomes a **super red giant**.
- When the core collapses to form a white dwarf, its mass is greater than 1.4 solar masses. The gravitational pressures are enormous and overcome the Fermi pressure. The electrons within the core combine with the protons to produce neutrons and neutrinos. The neutrinos escape and the central core becomes entirely packed with neutrons.
- The outer shells surrounding the neutron core rapidly collapse and rebound against the solid neutron core. This generates a shock wave, which explodes the surface layers of the star as a **supernova**.
- The supernova blasts off heavier elements like iron and oxygen and into space. It is worth remembering that all the elements on the Earth once originated from distant supernovae.
- For stars with mass in the range 3–10 solar masses, the remnant core is a **neutron star**. For stars with mass above 10 solar masses, the neutron core continues to collapse even further, resulting in a **black hole**. The gravitational field of a black hole is so strong that even light passing within a few kilometres cannot escape.

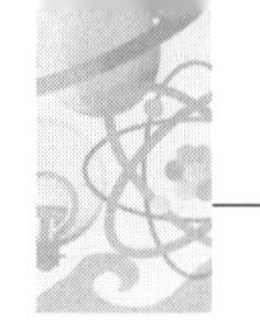

Measuring vast distances

For convenience, the distances within our own solar system are measured in astronomical units (AU).

An astronomical unit is the mean distance between the Earth and the Sun.

$1\ \text{AU} \approx 1.5 \times 10^{11}\ \text{m}$

The distance to stars and galaxies can be measured in light-years (ly).

The light-year is the distance travelled by light in a vacuum in a period of 1 year.

$1\ \text{ly} \approx 9.5 \times 10^{15}\ \text{m}$

The distance to stars and galaxies can also be measured in parsecs (pc). Before details of this can be given, you need to understand stellar parallax and the arc second.

- **Stellar parallax**: a nearby star A is photographed against the distant background of stars when the Earth is in its orbit at position X. Another photograph is taken exactly 6 months later when the Earth is at position Y. The star shows an apparent shifting of its position against the starry background — this is known as stellar parallax. However, to confuse matters, the term 'parallax', p, is used to denote half of the total angle subtended by the star between positions X and Y of the Earth.
- **Arc seconds**: the parallax p for all stars is significantly smaller than 1 degree. Very small angles are measured in arc seconds (or simply arc sec), where:

  $1\ \text{arc second} = \frac{1}{3600}\ \text{degrees}$

The parsec is best defined with the aid of a diagram (see below).

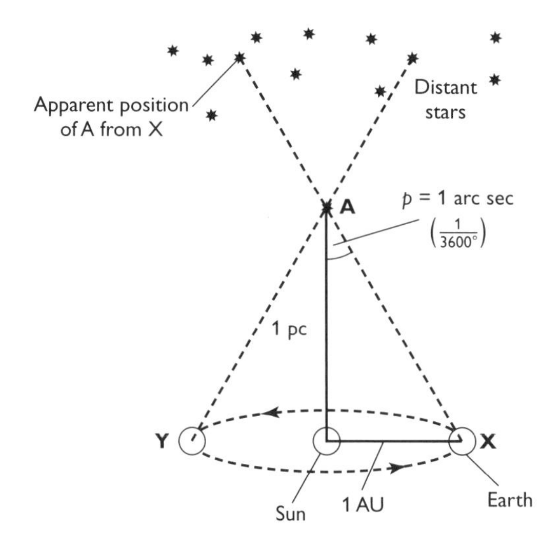

The parsec is defined as the distance that gives a parallax angle of 1 second of arc.

1 pc ≈ 3.1 × 10^{16} m

The parsec is short for parallax and second. Note:

$$\tan\left(\frac{1}{3600}\right) = \frac{1\,\text{AU}}{1\,\text{pc}}, \text{ hence } 1\text{ pc} = \frac{1.5\times10^{11}}{\tan(1/3600)} \approx 3.1\times10^{16}\text{ m}$$

A useful relationship between distance d in pc and parallax p in arc seconds is:

$$\boldsymbol{d = \frac{1}{p}}$$

Worked example

Sirius is a very bright star in the night sky. It has a parallax of 0.38 arc sec. Calculate its distance in parsecs and in light-years.

Answer

The distance in pc can be found using $d = \frac{1}{p}$. Therefore:

$$d = \frac{1}{0.38} = 2.63\text{ pc}$$

Since 1 pc ≈ 3.1 × 10^{16} m and 1 ly ≈ 9.5 × 10^{15} m, we have:

$$\text{distance of Sirius in ly} = 2.63 \times \frac{3.1\times10^{16}}{9.5\times10^{15}} \approx 8.6\text{ ly}$$

n Note: the light from Sirius will take about 8.6 years to reach us.

Olbers' paradox

The statement below is known as (Heinrich) Olbers' paradox:

For an infinite, uniform and static universe, the night sky should be bright because of light received from stars in all directions.

The following logical steps support this paradox:

- in an infinite universe, the number of stars in a spherical shell increases with distance^2
- the light received from each star decreases with distance^2
- these two factors cancel out and hence even at night, the sky should be bright

However, regular observation shows that the night sky is dark!

Like other paradoxes, Olbers' paradox is based on incorrect assumptions. The universe is neither static nor infinite. The fabric of space, which includes the galaxies, has been expanding since the (hot) big bang some 12 billion years ago.

The reasons for the night sky being dark are:

- the universe is finite in size
- the universe is not static but expanding (as confirmed by the redshift of light from distant galaxies and Hubble's law)
- the finite age of the universe means that light from distant galaxies has not yet reached us

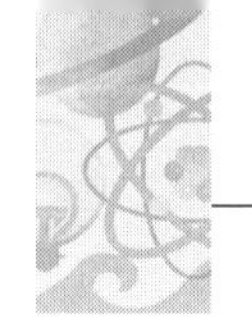

Redshift

You have already met the Doppler effect on p. 65. The Doppler effect can also be observed in the starlight from distant galaxies.

Consider light emitted by a star moving away from us. According to the Doppler effect, the wavelength of the measured light is longer. In fact, the *entire spectrum* from a star is shifted by the same fraction to longer wavelengths — this is known as **redshift**.

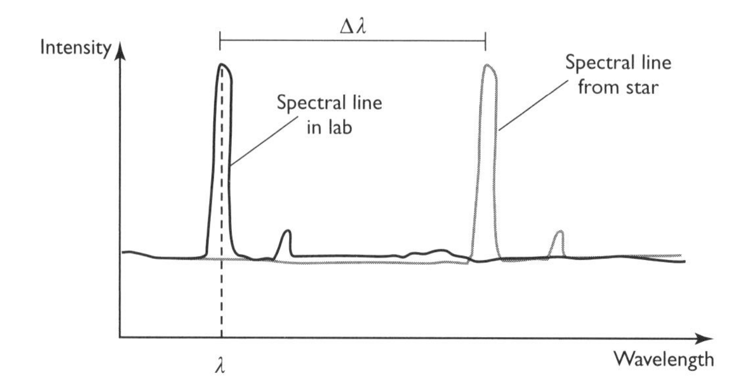

If a star is moving towards us, the entire spectrum is shifted to shorter wavelengths — this is known as **blueshift**.

The speed v of a star or a galaxy can be determined by measuring the wavelength of known spectral lines using the **Doppler equation** below:

$$\frac{\Delta\lambda}{\lambda} = \frac{v}{c}$$

where $\Delta\lambda$ is the change in the wavelength of a particular spectral line, λ is the wavelength measured in the laboratory and c is the speed of light.

Worked example

The wavelength of a particular spectral line in the laboratory is 119.5 nm. The same spectral line emitted from a star has a wavelength of 121.6 nm. Calculate the speed of the star and state whether it is moving away from us or coming towards us.

Answer

$$\frac{\Delta\lambda}{\lambda} = \frac{v}{c}$$

$$\Delta\lambda = 121.6 - 119.5 = 2.1\,\text{nm}$$

$$v = \frac{\Delta\lambda \times c}{\lambda} = \frac{2.1 \times 3.0 \times 10^8}{119.5} \approx 5.3 \times 10^6\ \text{m s}^{-1}$$

The wavelength of the light from the star is longer, hence it must be moving away (redshift).

Hubble's law

All galaxies are moving away from each other because the whole fabric of space is expanding. Hubble's law combines the distance x from us of a galaxy to its recessional speed v.

Hubble's law: the recessional speed of a galaxy is directly proportional to its distance from us.

According to Hubble's law:

$$v \propto x$$

or

$$\boldsymbol{v = H_0 x}$$

where H_0 is known as the Hubble constant.

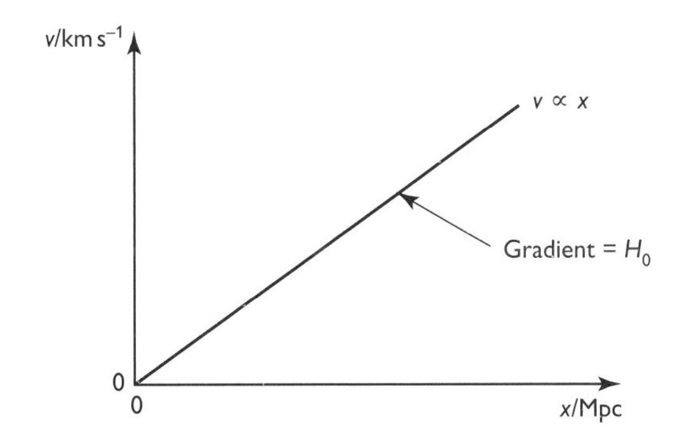

- The *gradient* of the velocity against distance graph is equal to the Hubble constant.
- The Hubble constant is often quoted in the unit 'km s^{-1} Mpc^{-1}' (kilometre per second per mega parsec). The SI unit for the Hubble constant is s^{-1}.
- The age of the universe is related to the Hubble constant as follows:

$$\textbf{age of universe} = \frac{1}{H_0}$$

Worked example

A galaxy at a distance of 100 Mpc has a recessional speed of 8000 km s^{-1}. Use this information to determine the Hubble constant in s^{-1} and estimate the age of the universe in years.

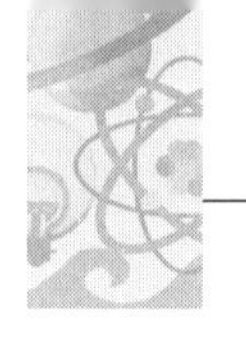

Answer

1 pc = 3.1×10^{16} m and 1 km s^{-1} = 10^3 m s^{-1}

$v = H_0 x$

Therefore

$$H_0 = \frac{v}{x} = \frac{8000 \times 10^3}{100 \times (10^6 \times 3.1 \times 10^{16})}$$

$$H_0 = 2.58 \times 10^{-18}\ s^{-1}$$

$$\text{age of universe} = \frac{1}{H_0} = \frac{1}{2.58 \times 10^{-18}} = 3.88 \times 10^{17}\ s$$

According to the *Data, Formulae and Relationship* booklet, 1 year = 3.16×10^7 s, therefore:

$$\text{age of universe} \approx \frac{3.88 \times 10^{17}}{3.16 \times 10^7} = 12 \times 10^9 \text{ years (12 billion years)}$$

The evolution of the universe

Cosmological principle

This important principle has three elements.

(1) The universe is *homogeneous*. On a large scale, its density is the same everywhere.

(2) The universe is *isotropic*. This means that the universe is the same in all directions. The evidence for this comes from the background microwave radiation (see p. 73).

(3) The laws of physics are *universal*. This implies that all laws of physics can be applied everywhere.

Evolution of the universe

The **big bang model**, also known as the **standard model** of the universe, assumes that space and time evolved from a *singularity* in an event that took place about 12 billions years ago. This is how the universe has evolved.

- The big bang took place about 12 billion years ago. The universe was infinitesimally small, infinitely dense and extremely hot. All four forces of nature (gravitational, electromagnetic, strong nuclear and weak nuclear) were unified.
- The expansion of the universe led to its cooling.
- After about 10^{-6} s, the temperature of the universe was about 10^{14} K. The universe consisted of energetic quarks and leptons.
- After about 10^{-3} s, the temperature of the universe was about 10^{12} K. The strong nuclear force became dominant and combined the quarks to form hadrons, including protons and neutrons.

- At a temperature of about 10^7 K, fusion reactions between protons produced a significant amount of helium nuclei. (This helium is known as 'primordial helium'. About 25% of the matter in the observable universe is helium.)
- At a temperature of about 10^4 K, electrons combined with nuclei to form hydrogen and helium atoms.
- Gravitational forces became dominant. Hydrogen and helium clumped together to form stars and eventually clusters of galaxies.
- The temperature of the universe was now 2.7 K. At this temperature, the universe is saturated with electromagnetic waves in the microwave region of the spectrum. On the Earth, we can observe this **background microwave radiation** as having the same intensity in all directions (isotropic).

Evidence for the big bang

The following is a list of the main observational evidence for the big bang model.

(1) The universe is *expanding*.

(2) *Hubble's law* shows that all galaxies are receding from us.

(3) The temperature of the universe is 2.7 K, with small *'ripples'* of $\pm 10^{-5}$ K. (The ripples were crucial in the clumping of matter and the eventual formation of stars and galaxies.)

(4) The universe is saturated with background microwave radiation.

(5) The most distant galaxies, and hence the youngest, show chemical composition of 25% primordial helium.

Critical density

The final fate of the universe depends on the density of the universe. This density is difficult to determine because astronomers cannot observe all the matter in the universe — either because the matter does not emit much electromagnetic radiation or does not interact easily because it is electrically neutral. The mass of the universe is not just that due to galaxies, but is also due to the existence of *dark matter*. The most likely candidates for dark matter are neutrinos and black holes.

The fate of the universe depends on its density compared with the critical density. The critical density ρ_0 is given by the equation:

$$\rho_0 = \frac{3H_0^2}{8\pi G}$$

where G is the gravitational constant, $6.67 \times 10^{-11}\,\mathrm{N\,m^2\,kg^{-2}}$, and H_0 is the Hubble constant, in $\mathrm{s^{-1}}$.

What is the critical density if we assume a value of $2.6 \times 10^{-18}\,\mathrm{s^{-1}}$ for the Hubble constant?

$$\rho_0 = \frac{3 \times (2.6 \times 10^{-18})^2}{8\pi \times 6.67 \times 10^{-11}} \approx 1.2 \times 10^{-26}\ \mathrm{kg\,m^{-3}}$$

This is an extremely small density compared with normal matter; for example, water has density of $10^3\,\mathrm{kg\,m^{-3}}$. The critical density is equivalent to about 7 protons per cubic metre.

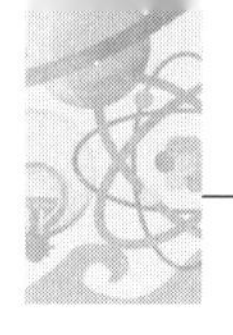

The fate of the universe

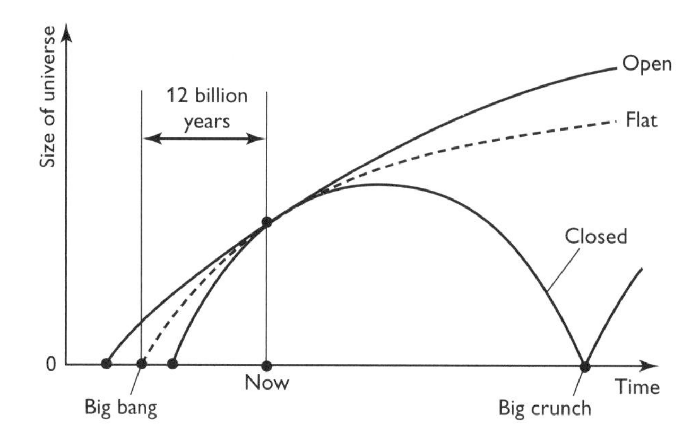

The evolution of the universe can be summarised as follows.

- The universe will be **closed** if its density is *greater* than the critical density. The gravitational force between matter (galaxies, dark matter etc.) in such a universe is strong enough to decelerate and *halt* the expansion of the universe. The universe will then contract, getting hotter and hotter as it approaches the **big crunch**. In this model, the universe could oscillate between big bangs and big crunches.
- The universe will be **open** if its density is *less* than the critical density. The gravitational force between matter in such a universe is too weak to decelerate or stop the expansion of the universe. The universe will expand *forever*, getting colder and colder as its temperature approaches absolute zero. This is indeed a bleak end.
- The universe will be **flat** if its density is *equal* to the critical density. The rate of expansion of such a universe tends to zero as the volume of the universe tends towards a certain limit. Most cosmologists believe that our universe is flat.

Questions & Answers

This section contains questions similar in style to those you can expect in the exam paper for Unit G485. The responses of two candidates are given. The answers from Candidate A are similar to those expected from a student who has reached grade-A standard in this unit. The answers given by Candidate B are typical of those of a grade-C candidate and may be incomplete or inappropriate. Answers from weaker candidates than this are not very instructive at this stage, because there are either too many questions unanswered or the answers show a lack of knowledge of basic physics.

Questions at A2 look for connections between different topics and also have synoptic elements — this is reflected in the questions chosen for this section. The Question and Answer section can be used in several ways. The best way to engage your brain is to write something on a piece of paper. There is no value at all in just reading through the questions and answers given by the two candidates. Here are some possible strategies. You could:

- Attempt a question yourself and then mark it using the comments from the examiner. This will enable you to compare the responses of two candidates and learn from your mistakes.
- Mark the answers given by the two candidates and then see whether you agree with the mark awarded by the examiner.

Examiner's comments

Every mark scored by a candidate is shown by a tick at the relevant place. For most part, the total number of ticks (✓) and crosses (✗) should add up to the total mark for the question. All candidates' responses are followed by examiner's comments. These are denoted by the icon e. These comments focus on the errors made by candidates and sometimes offers alternative ways of securing the marks. The comments at the end of each question provide valuable tips for answering examination questions.

Question 1

(a) Define *capacitance*. (1 mark)

(b) Two parallel metal plates have a 12 mm separation. This arrangement has a capacitance of 7.4 pF. The plates are connected to a 5.0 kV supply.

(i) Calculate the charge on the positive plate. (2 marks)

(ii) Calculate the electric field strength between the plates. (2 marks)

(iii) The diagram below shows the path of an electron beam between the charged parallel plates.

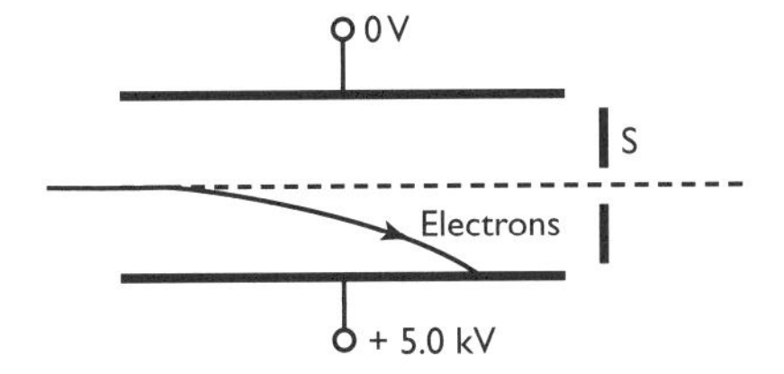

Describe and explain how a magnetic field may be used to make the electron beam travel in a straight line and pass though the slit S. (3 marks)

Total: 8 marks

Candidates' answers to Question 1

Candidate A

(a) Capacitance is the charge per unit potential difference. ✓

Candidate B

(a) $C = \frac{Q}{V}$, where C is the capacitance, Q is the charge on one of the plates and V is the voltage across the capacitor. ✓

Both candidates have made a good start by correctly defining capacitance.

Candidate A

(b) (i) $Q = VC = 5000 \times 7.4 \times 10^{-12} = 3.7 \times 10^{-8}$ C ✓✓

Candidate B

(b) (i) charge = $VC = 5000 \times 7.4 \times 10^{-9}$ ✗

charge = 3.7×10^{-5} C ✓ (*error carried forward*)

Candidate A shows good knowledge of prefixes. Sadly, Candidate B uses a factor of 10^{-9} instead of 10^{-12} for pico. However, the examiner would award 1 mark for the calculation by applying the 'error carried forward' rule.

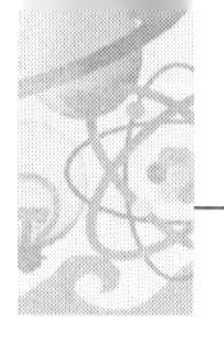

Candidate A

(b) (ii) $E = \frac{V}{d} = \frac{5000}{12 \times 10^{-3}} = 4.17 \times 10^5\ \text{V m}^{-1}$ ✓✓

Candidate B

(b) (ii) $E = \frac{V}{d} = 4.2 \times 10^5\ \text{N C}^{-1}$ ✓✓

Candidate A has provided a perfect solution with clear substitution and the correct unit for electric field strength. Candidate B has quoted the right answer to 2 s.f., but has been a bit wayward by not showing the calculation. The alternative unit, N C^{-1}, is correct.

Candidate A

(b) (iii) Each electron will travel in a straight line when the magnetic force is equal but opposite to the electric force experienced by the electron. ✓
Therefore: $BQv = QE$ ✓
The direction of magnetic field must be into the plane of the paper. ✓

Candidate B

(b) (iii) The magnetic field can be applied at right angles to the electric field. ✓

Candidate A has shown excellent understanding of 'crossed' electric and magnetic fields. Candidate B has been too brief and should have tried to write a little more for this 3-mark question.

The presentation of Candidate A's answer is immaculate and the examiner would have no problems following the answers and awarding full marks. Candidate B has picked up 5 marks but has lost some easy marks. The answer from this candidate was too brief for part (b)(iii). Just gaining 1 extra mark would have elevated Candidate B to a grade B.

Question 2

(a) **Define the terms *magnetic flux* and *magnetic flux linkage*.** (2 marks)

(b) **State Faraday's law of electromagnetic induction.** (1 mark)

(c) **The diagram below shows a coil of cross-sectional area 3.6×10^{-4} m^2 and having 1200 turns placed in a uniform magnetic field of flux density 0.13 T.**

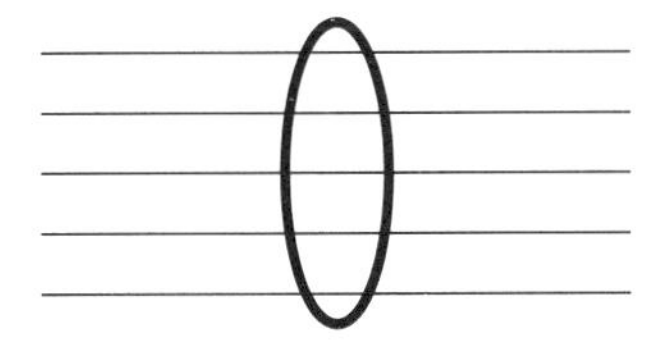

The plane of the coil is at right angles to the direction of the magnetic field. The magnetic field is provided by a strong electromagnet.

(i) **Explain why there is no induced e.m.f. across the ends of the coil when it is moved in a direction to the magnetic field.** (1 mark)

(ii) **The magnetic field is switched off. The magnetic flux density collapses to zero in a time of 3.0 ms. Calculate the induced e.m.f. across the ends of the coil.** (4 marks)

Total: 8 marks

Candidates' answers to Question 2

Candidate A

(a) When the magnetic field is normal to the plane of the circuit, then:
magnetic flux = magnetic flux density × cross-sectional area ✓
The flux linkage depends on the number of turns and is given by:
magnetic flux linkage = number of turns × magnetic flux ✓

Candidate B

(a) $\phi = BA$, where ϕ is the flux, B is the flux density and A is the area of the circuit. ✓
magnetic flux linkage = $N\phi$, where N is the number of turns ✓

e Both candidates have made a decent start and have shown good recall skills. Candidate B's answer for magnetic flux linkage is brief, but all the terms have been defined.

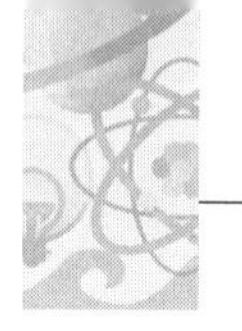

Candidate A

(b) According to Faraday's law, the size of the induced e.m.f. is equal to the rate of change of magnetic flux linkage. ✓

Candidate B

(b) $E = \dfrac{\Delta(N\phi)}{\Delta t}$

where E is the induced e.m.f. in a circuit, $\Delta(N\phi)$ is the change in magnetic flux linkage and Δt is the change in time. ✓

These answers are equally acceptable. Candidate A has defined the law using words and Candidate B has opted for an equation.

Candidate A

(c)(i) An e.m.f. can only be produced if the magnetic flux changes. Since the flux linkage does not change, there is no induced e.m.f. across the ends of the coil. ✓

Candidate B

(c)(i) To get an e.m.f. from coil, the strength of the magnetic field or the size of its cross-sectional area must change. Since neither happens, the e.m.f. is zero. ✓

Both candidates have demonstrated a good understanding of Faraday's law.

Candidate A

(c)(ii) induced e.m.f. = rate of change of magnetic flux

$$= \frac{\Delta(NBA)}{\Delta t} = \frac{1200 \times 3.6 \times 10^{-4} \times 0.13}{0.03} \checkmark\checkmark$$

$$= 18.7 \text{ volts} \checkmark\checkmark$$

Candidate B

(c)(ii) e.m.f. = 0.13/0.003 = 43 ✗

Candidate A has clearly set out the answer. The final magnetic flux is zero and hence the change in the flux is the same as the initial flux through the coil. In this question there is 1 mark for writing the correct unit for e.m.f. (volts). Candidate B has simply resorted to guessing and has wrongly divided the magnetic flux density by the time.

Overall, Candidate A has scored maximum marks and Candidate B has scored 5 marks (out of 8). Candidate B managed to quote Faraday's law in (b), but then failed to use the data in (c) to solve the problem. Just picking up 1 more mark (e.g. writing the unit) would have lifted Candidate B's performance to a higher grade.

Question 3

(a) **Explain the terms *hadrons*, *leptons* and *quarks*.** (3 marks)

(b) **The reaction below shows what happens inside a nucleus during a beta-minus decay.**

$${}^{1}_{0}n \rightarrow {}^{1}_{1}p + {}^{0}_{-1}e + \bar{v}$$

(i) **The reaction shows a neutron decaying into a proton. Identify the other two particles.** (1 mark)

(ii) **In this reaction, the mass decreases by 8.4×10^{-4} u. Calculate the total energy released as kinetic energy in this reaction.** (3 marks)

(c) **The mean separation between the protons in a helium nucleus is about 10^{-15} m. Determine the ratio of the repulsive electrostatic force and attractive gravitational force experienced by the protons. Hence, explain how this ratio suggests the existence of another force between the protons.** (4 marks)

Total: 11 marks

Candidates' answers to Question 3

Candidate A

(a) Hadrons are particles (protons and neutrons) that experience the strong nuclear force. ✓
Leptons are group of particles (e.g. electron) that do not feel the strong nuclear force. ✓
Quarks are the fundamental particles that make up the hadrons. ✓

Candidate B

(a) Hadrons feel the strong nuclear force and leptons do not. ✓✓
Quarks such as 'up' and 'down' are inside all hadrons. ✓

This is a comprehensive answer from Candidate A. Candidate B has been brief, but has managed to convey the important points to the examiner. This is a good start by both candidates.

Candidate A

(b) **(i)** The other two particles are an electron and an antineutrino. ✓

Candidate B

(b) **(i)** The last particle is a neutrino and the other is an electron. ✗

Candidate A continues to do well by providing a succinct answer. Candidate B has not identified the antineutrino – the bar above the 'v' implies antimatter.

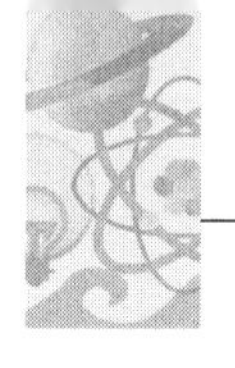

Candidate A

(b) (ii) According to Einstein's mass–energy equation, the energy released is given by:
energy = Δmc^2 = $(8.4 \times 10^{-4} \times 1.66 \times 10^{-27}) \times (3.0 \times 10^8)^2$ ✓✓
energy = 1.25×10^{-13} joules ✓

Candidate B

(b) (ii) $\Delta E = \Delta mc^2 = 8.4 \times 10^{-4} \times (3.0 \times 10^8)^2 \approx 7.6 \times 10^{13}$ J ✓✗✗

Candidate A has a good understanding of the mass–energy equation and realised that the change in mass had to be converted into kilograms. Sadly, Candidate B knew what equation to use, but made a fundamental mistake by using the mass in unified atomic mass units (u). The examiner has awarded 1 mark for using the mass–energy equation.

Candidate A

(c) ratio = $\dfrac{e^2}{4\pi\varepsilon_0 r^2} \div \dfrac{Gm^2}{r^2} = \dfrac{e^2}{4\pi\varepsilon_0 Gm^2}$ ✓

ratio = $\dfrac{(1.6 \times 10^{-19})^2}{4\pi \times 8.85 \times 10^{-12} \times 6.67 \times 10^{-11} \times (1.67 \times 10^{-27})^2}$ ✓

ratio = 1.2×10^{36} ✓

The repulsive force between the protons is much much greater than the attractive gravitational force. There must be another attractive force between the protons — the strong nuclear force. ✓

Candidate B

(c) gravitational force = $\dfrac{GMm}{r^2} = \dfrac{6.67 \times 10^{-11} \times (1.67 \times 10^{-27})^2}{(10^{-15})^2} = 1.86 \times 10^{-34}$ N ✓

electrical force = $\dfrac{Qq}{4\pi\varepsilon_0 r^2} = \dfrac{(1.6 \times 10^{-19})^2}{4 \times 3.142 \times 8.85 \times 10^{-12} \times (1 \times 10^{-15})^2} = 230$ N ✓

The ratio is 1.86×10^{-34} : 230. ✗

The gravitational force cannot hold the protons but the strong nuclear (attractive) force can. ✓

Candidate A has opted for an algebraic approach – it is interesting how the ratio is independent of the actual separation between the protons. Candidate B worked out the individual forces but failed to determine the actual ratio between the forces. Both candidates correctly identified the strong nuclear force as the attractive force binding the protons together.

Candidate A typifies the capabilities of a high-scoring candidate. The answers contain a good blend of physics and mathematics. The knowledge base of this candidate is good enough to score an A* grade. Candidate B scored 7 marks (out of 11) and could have easily picked up two extra marks in part (b) (ii) by converting the mass into kilograms. Sloppy work has cost Candidate B at least one whole grade.

Question 4

(a) Technetium-99m is a radioactive isotope with half-life of 6.0 hours. A particular sample of technetium has an activity of 300 MBq.

(i) Show that the sample has about 9×10^{12} active nuclei of technetium. (2 marks)

(ii) Estimate the total mass of technetium in the sample. (2 marks)

(iii) Determine the activity of the sample after a time of a day. (2 marks)

(b) State one of the medical uses of technetium-99m. (1 mark)

(c) Name the main components of a gamma camera. Briefly describe the function of each component. (4 marks)

Total: 11 marks

■ ■ ■

Candidates' answers to Question 4

Candidate A

(a) (i) $A = \lambda N$ and $\lambda = 0.693/t_{1/2}$

$\lambda = 0.693/(6.0 \times 3600) = 3.21 \times 10^{-5}\,s^{-1}$ ✓

number of nuclei, $N = \dfrac{300 \times 10^6}{3.21 \times 10^{-5}} = 9.35 \times 10^{12}$ ✓

Candidate B

(a) (i) decay constant $= \dfrac{0.693}{2.14 \times 10^4} = 3.2 \times 10^{-5}$ per second ✓

activity = decay constant × number of nuclei

number of nuclei = activity ÷ decay constant $= 300 \times 10^6 \div 3.2 \times 10^{-5}$

$\approx 9 \times 10^{12}$ ✓

Both candidates have gained full marks. Candidate B should have quoted the final answer to more significant figures in this 'show' question.

Candidate A

(a) (ii) molar mass = 99 g

mass $= \dfrac{9.35 \times 10^{12}}{6.02 \times 10^{23}} \times 99 = 1.54 \times 10^{-9}$ grams ✓✓

Candidate B

(a) (ii) mass $= 9 \times 10^{12} \times 1.66 \times 10^{-27} = 1.49 \times 10^{-14}$ kg ✓✗

Candidates often find such questions very tough. It requires synoptic understanding of moles and the Avogadro constant. Candidate A has gained both marks. Candidate B could also have scored full marks by multiplying the final answer by a factor of 99; there are 99 nucleons in an isotope of Tc-99m. The examiner has been kind to award 1 mark.

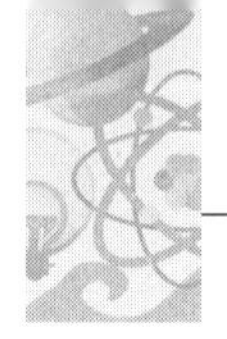

Candidate A

(a) (iii) A day has 4 half-lives. ✓

Therefore, the activity = 300 MBq × $\left(\frac{1}{2}\right)^4$ = 18.8 MBq ✓

Candidate B

(a) (iii) 24 hours = 1 day = 4 half-lives ✓

activity = 300/16 = 19 MBq ✓

Candidates A and B have both provided good answers, with both realising that there are four lots of 6 hours in a day.

Candidate A

(b) Tc-99m is used to diagnose the function of the brain. ✓

Candidate B

(b) Technetium can be used to check the function of the lungs using a gamma camera. ✓

Technetium is a versatile tracer; both candidates have correctly identified one use.

Candidate A

(c) A gamma camera has the following components:

- Collimator — made from lead tubes that only allows gamma photons coming directly a certain direction. ✓
- Scintillator — this changes gamma photons into increased number of visible light photons. ✓
- Photomultiplier tubes — the visible light produces many electrons via the photoelectric effect. Each photon of light produces a single electrical pulse from the tube. ✓
- Computer — this collects all the pulses from the photomultiplier tubes and forms a composite image of the patient's organ. ✓

Candidate B

(c) The components are: computer, photomultiplier (tubes) and collimator.

The computer is the one that gets all the signals from the photomultipliers to form a picture of the patient's organ. ✓

Candidate A's answer is immaculate. Each component is named and its function clearly stated. Candidate B has missed the scintillator and has not described the function of most of the components.

Overall, Candidate A has scored maximum marks and is definitely on target for an A* grade. Candidate B has scored 7 marks (out of 11) and has missed too many opportunities. The answer to part (c) was incomplete and lacked rigour — the command word 'describe' in part (c) failed to trigger the right response from this candidate. Candidate B needs to practise answering long-answer questions.

Question 5

(a) Briefly describe how X-rays are produced in an X-ray tube. State one of the properties of X-rays. (4 marks)

(b) Describe the three main absorption (attenuation) mechanisms when X-rays travel through matter. (6 marks)

(c) An X-ray beam of intensity 20 W m^{-2} and cross-sectional area 5.0×10^{-3} m^2 is incident on bone. The graph below shows the variation of the intensity *I* of the X-ray beam with the thickness *x* of bone.

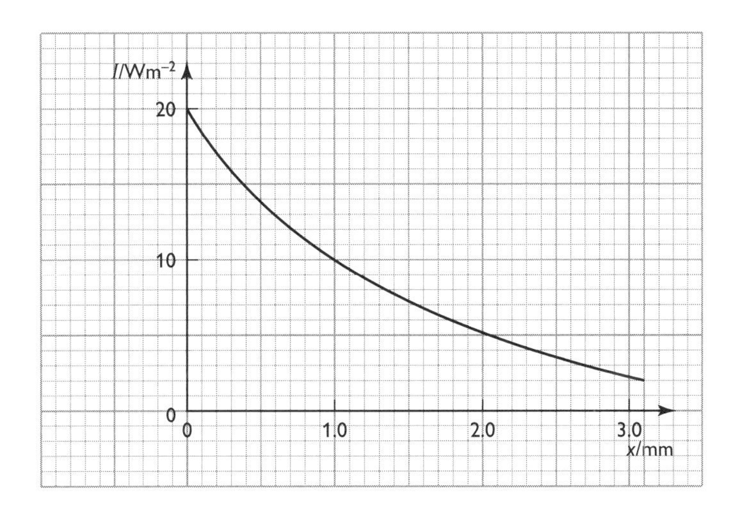

Calculate:

(i) the incident power incident on the bone (2 marks)

(ii) the attenuation (absorption) coefficient of bone (3 marks)

Total: 15 marks

Candidates' answers to Question 5

Candidate A

(a) Electrons are produced by a hot filament (cathode). ✓ These are accelerated towards a target metal held at high positive voltage. ✓ When the electrons are stopped by the metal (e.g. tungsten), their kinetic energy is transformed into X-ray photons. ✓
Property: X-rays travel at the speed of light (3×10^8 m s^{-1} in a vacuum). ✓

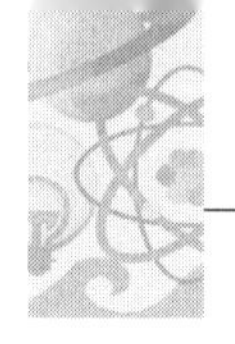

Candidate B

(a) One property is that X-rays are transverse electromagnetic waves. ✓
Electrons are 'boiled off' a hot filament. ✓ A high voltage is applied between the tungsten target metal and the filament. ✓ The kinetic energy of the electrons is converted to X-rays when they are stopped. ✓

Both candidates have done extremely well with their description of how X-rays are produced.

Candidate A

(b) Photoelectric effect — this is when the photon disappears and the energy is used to remove one of the orbiting electrons of the target metal. ✓✓
Pair production — a high-energy photon (>1.02 MeV) disappears and materialises as an electron–positron pair. ✓✓
Compton effect — the photon is scattered in a different direction by an atomic electron. The electron is removed and the scattered photon has lower frequency or energy. ✓✓

Candidate B

(b) The mechanisms are: Compton, photoelectric and pair production. ✓✓✓

Candidate A has named the mechanisms and described each mechanism. Candidate B has answered only half of the question. There is no explanation or description of the mechanisms – 3 valuable marks have been lost for not reading the question with care.

Candidate A

(c) (i) intensity = power/cross-sectional area
power = $I \times A = 20 \times 5.0 \times 10^{-3} = 0.1$ watts ✓✓

Candidate B

(c) (i) I can get the power by multiplying the 'W m^{-2}' by 'm^2'.
Therefore, $P = 20 \times 0.005 = 0.1$ W ✓✓

A perfect answer from Candidate A. Candidate B has used the units to work out what to do; there is nothing wrong with this approach. The important thing is that the answer and the unit for the power are correct.

Candidate A

(c) (ii) $I = I_0 e^{-\mu x}$
From the graph $I_0 = 20\,\text{W m}^{-2}$, $I = 10\,\text{W m}^{-2}$ when $x \approx 1.0$ mm ✓
$10 = 20e^{-\mu \times 1.0}$ ✓
$\ln(0.5) = -\mu$
$\mu \approx 0.70$ mm ✓

Candidate B

(c) (ii) I don't know what is needed here — sorry! ✗✗✗

This is a tough but fair question at A2. Candidate A's answer is exemplary. Candidate B sadly has not attempted to answer the question. Grade-C candidates often have serious gaps in their knowledge.

Candidate A has gained full marks and showed excellent comprehension of X-rays. Candidate B just managed to get a grade C by picking up 9 marks (out of 15). Candidate B should have read part (b) more carefully. Sadly, this candidate made no use of the equation $I = I_0e^{-\mu x}$, even though it is given in the *Data, Formulae and Relationship* booklet — just substituting values into this equation would have given a couple of extra marks.

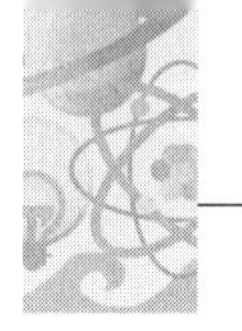

Question 6

(a) Explain how an ultrasound transducer can emit and detect ultrasound. (2 marks)

(b) Define the *acoustic impedance* of a material. (1 mark)

(c) Explain the term *impedance matching* in the context of an ultrasound scan. (3 marks)

(d) The acoustic impedance of muscle and soft tissue is 1.7×10^6 kg m^{-2} s^{-1} and 1.3×10^6 kg m^{-2} s^{-1} respectively. Calculate the fraction of ultrasound intensity transmitted through the boundary between muscle and soft tissue. (3 marks)

Total: 9 marks

Candidates' answers to Question 6

Candidate A

(a) A piezoelectric transducer has a thin film of piezoelectric crystal.

The crystal emits ultrasound when connected to an alternating voltage because it vibrates at the same frequency as the applied signal. ✓

The crystal has an e.m.f. induced when ultrasound hits it. ✓

Candidate B

(a) The transducer works on the principle of the piezoelectric effect.

When the transducer material (crystal) is vibrated by the ultrasound, it generates an e.m.f. across its ends. ✓

The same material can be made to emit high-frequency sound by applying an alternating voltage — this makes it vibrate. ✓

e Both candidates have demonstrated a good understanding of the piezoelectric effect.

Candidate A

(b) acoustic impedance = density of material × speed of ultrasound ✓

Candidate B

(b) $Z = \rho c$

where Z = acoustic impedance, ρ = density and c = speed of sound ✓

e Both candidates have been awarded the 1 mark, with Candidate A writing a clear word equation and Candidate B writing the symbol equation and then defining all the terms.

Candidate A

(c) When a transducer is placed directly on skin, most of the ultrasound is reflected back. ✓ This is a problem because an ultrasound scan image will not be very bright or clear. So a special gel is smeared on the skin. ✓ The acoustic impedance of the gel is almost the same as that of skin — hence most of the ultrasound intensity is transmitted into the patient. ✓

Candidate B

(c) For acoustic matching, a gel is applied between the skin and the transducer. ✓ This helps to produce a good scan.

Both candidates have identified the use a special gel. However, only Candidate A has gone on to explain what it does in terms of minimising the amount of reflected ultrasound. Candidate B's answer is too brief for the 3-mark question.

Candidate A

(d) $\frac{I_r}{I_0} = \left(\frac{Z_2 - Z_1}{Z_2 + Z_1}\right)^2 = \left(\frac{1.7 - 1.3}{1.7 + 1.3}\right)^2 = 0.018\ (1.8\%)$ ✓✓

Hence the fraction transmitted is:

$1 - 0.018 = 0.982$ ✓ (or 98.2%)

Candidate B

(d) $\frac{I_r}{I_0} = \frac{(Z_2 - Z_1)^2}{(Z_2 + Z_1)^2} = \frac{(1.3 \times 10^6 - 1.7 \times 10^6)^2}{(1.3 \times 10^6 + 1.7 \times 10^6)^2} = 0.13$ ✓✗

fraction transmitted $= 1 - 0.13 = 0.87$ ✓ (*error carried forward*)

Candidate A has produced a clear and correct answer. Candidate B has applied the correct equation but has forgotten to square the final answer. The answer from this candidate looks a bit clumsy because, unlike Candidate A, all the 10^6 factors have been written down.

Candidate A has produced immaculate answers and has shown a superb understanding of ultrasound. Candidate B has lost valuable marks in part (c) by not explaining the purpose of the gel. In part (d), Candidate B made a silly error by not squaring the final answer — this has cost the candidate at least one grade.

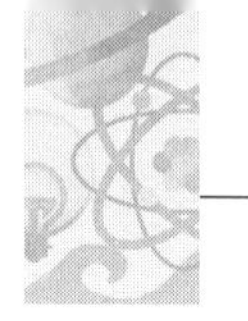

Question 7

(a) Explain how energy is produced in the core of a star. (4 marks)

(b) Explain the terms *supernova, white dwarf* and *black hole*. (6 marks)

(c) (i) The diagram below shows a binary star; the stars rotate about their centre of mass.

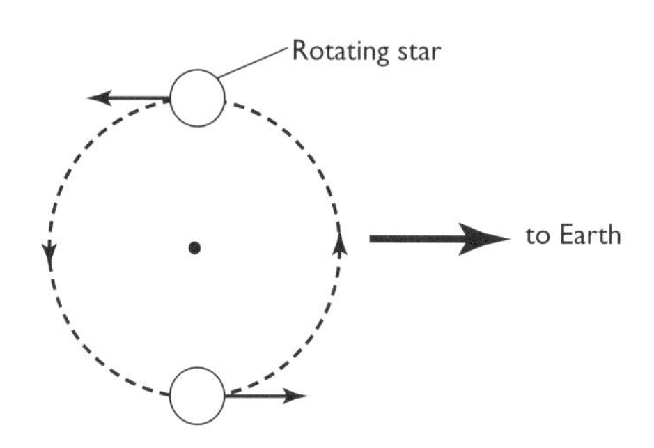

The light coming from the stars is analysed on the Earth. Explain how the Doppler effect can be used to show that the stars are rotating. (2 marks)

(ii) On the Earth, a particular spectral line from hydrogen has a wavelength of 410 nm. The same spectral line from a distant galaxy has a longer wavelength of 472 nm. Calculate the recessional speed of this galaxy. (3 marks)

Total: 15 marks

Candidates' answers to Question 7

Candidate A

(a) All stars generate energy through thermonuclear fusion reactions. ✓

A typical reaction is $4\,^{1}_{1}H \rightarrow \,^{4}_{2}He + 2\,^{0}_{+1}e + 2\nu$, which is known as 'hydrogen burning'. ✓

In such reactions, there is a reduction in mass equal to Δm. ✓ This mass is then changed into energy (kinetic, photons etc.) according to Einstein's mass–energy equation: $\Delta E = \Delta mc^2$. ✓

Candidate B

(a) Fusion reactions in stars create the energy. ✓ Such reactions fuse hydrogen nuclei to produce helium nuclei. ✓

Candidate A has carefully kept an eye on the available marks and has written a comprehensive account of nuclear fusion. Candidate B has been brief and sadly has lost 2 easy marks.

Candidate A

(b) Supernova is an event where the outer shells of a super massive star implode. ✓ This results in rapid collapse of the core and rebounding of the shells creates a shockwave which spills out hot stellar matter into space. ✓

A white dwarf is a remnant of a star with mass less than 3 solar masses. ✓ A white dwarf is a dead star because no fusion reactions take place. ✓ It is also very dense star ($10^{15}\,kg\,m^{-3}$).

A black hole is a remnant of a super massive star (mass > 10 solar masses). ✓ The gravitational field of a black hole is very strong and prevents light from escaping (this is why it is 'black' against the starry sky). ✓

Candidate B

(b) A supernova is an exploding star. ✗✗

A white dwarf is what is left after a low-mass star ejects a planetary nebula. ✓ A white dwarf does not produce any fusion — it leaks away photons from past reactions. ✓

A black hole is very small and may even have infinite density. ✓ It sucks in everything in space — even spaceships! ✗

This is another excellent answer from Candidate A and contains good physics. Candidate B has not given enough details for the supernova or the black hole. An 'exploding star' is too vague at A2 and examiners would expect better understanding.

Candidate A

(c) (i) The light from one of the stars is blue-shifted when coming towards us ✓ and then red-shifted when travelling away from us. ✓ This alternating blue and red shift shows rotation.

Candidate B

(c) (i) The star's spectrum will have longer wavelengths when it is moving away ✓ and the same spectrum will have shorter wavelengths when coming towards us. ✓

Both candidates have correctly answered the question. It is worth noting that even though the answers look different, the physics is still the same.

Candidate A

(c) (ii) $\frac{\Delta\lambda}{\lambda} = \frac{v}{c}$, $\Delta\lambda = 472 - 410 = 62\,\text{nm}$ and $\lambda = 410\,\text{nm}$

$$\frac{v}{3.0 \times 10^8} = \frac{62}{410} \quad ✓✓$$

speed $v = 4.5 \times 10^7\,m\,s^{-1}$ ✓

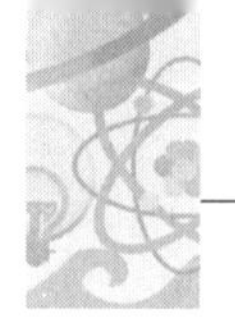

Candidate B

(c) (ii) $\frac{\Delta\lambda}{\lambda} = \frac{472 - 410}{410} = 0.151$ ✓

$\frac{v}{c} = \frac{\Delta\lambda}{\lambda} = 0.151$ ✓

$v = 0.151 \times 3.0 \times 10^8 = 4.54 \times 10^7$ m/s ✓

Both candidates have done well here. The key stages of the calculations are clearly set out by both candidates.

Overall, Candidate A has picked up maximum marks. The knowledge of this candidate is clear to see from the way the answers are represented. Candidate B has gained 10 marks (out of 15). This candidate had good analytical skills but fell down with the extended writing questions.

Question 8

(a) **State Hubble's law.** (1 mark)

(b) **The diagram shows a graph of the recessional speed v of our nearest galaxies against their distance x from the Earth.**

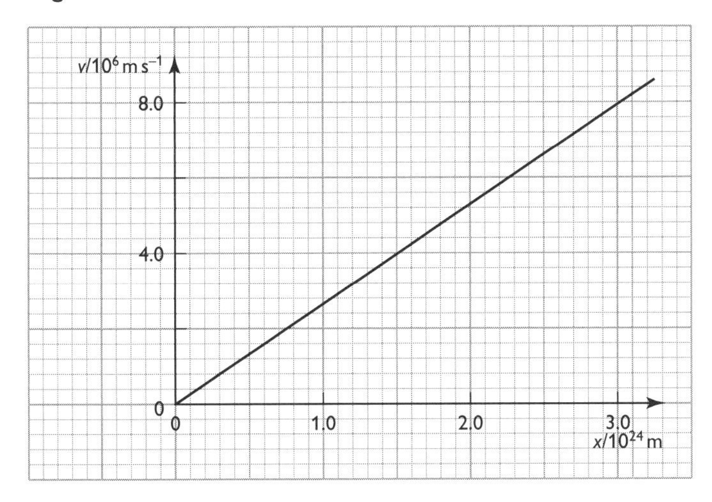

Use this graph to determine the Hubble constant and hence estimate the age of the universe in years. (4 marks)

(c) **(i)** **One of the strongest pieces of evidence for the big bang is Hubble's law and the recession of all galaxies. State two other pieces of observational evidence that support the big bang.** (2 marks)

(ii) **With the help of a labelled diagram, explain what is meant by a *closed universe*.** (3 marks)

Total: 10 marks

Candidates' answers to Question 8

Candidate A

(a) According to Hubble's law, the recessional speed of a galaxy is proportional to its distance from us. ✓

Candidate B

(a) Hubble's law: $v \propto x$, where x = distance of galaxy and v is its speed of recession. ✓

This is a decent start from both candidates. At A2, most candidates are expected to do well with this easy starter.

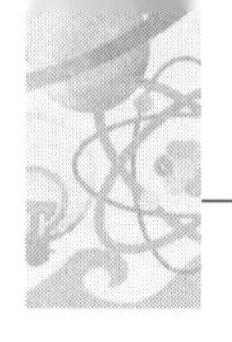

Candidate A

(b) gradient = Hubble constant, H_0

$H_0 = 8.0 \times 10^6/3.0 \times 10^{24} = 2.67 \times 10^{-18}\,s^{-1}$ ✓

age $= H_0^{-1} = (2.67 \times 10^{-18})^{-1} = 3.75 \times 10^{17}$ s ✓ ✓

This is about 12 billion years since the big bang. ✓

Candidate B

(b) age $= 1/H_0$

$$H_0 = \frac{4.0 \times 10^6}{1.5 \times 10^{24}} = 2.67 \times 10^{-18} \text{ per second}$$ ✓

age $= 3.8 \times 10^{17}$ seconds ✓ ✓

Candidate A has read the question carefully and given the age of the universe in years; sadly, the same cannot be said for Candidate B.

Candidate A

(c) (i) The expansion of the universe led to cooling. The temperature of the universe is about 2.7 K. ✓

At this temperature, the universe 'emits' background microwave radiation. ✓

Another evidence is the variation of temperature of the universe, about 10^{-5} K 'ripples'. There is also more matter than antimatter in our universe — as predicted by the standard model.

Candidate B

(c) (i) The temperature of space is 3 K. ✓ It is at this low temperature because expansion of space led to cooling.

Candidate A's answer is comprehensive; it has four correct statements (sadly, the examiner could not award more marks). Candidate B has been brief by giving just one piece of observational evidence. The examiner has allowed both 2.7 K and 3 K for the temperature of the universe.

Candidate A

(c) (ii) ✓

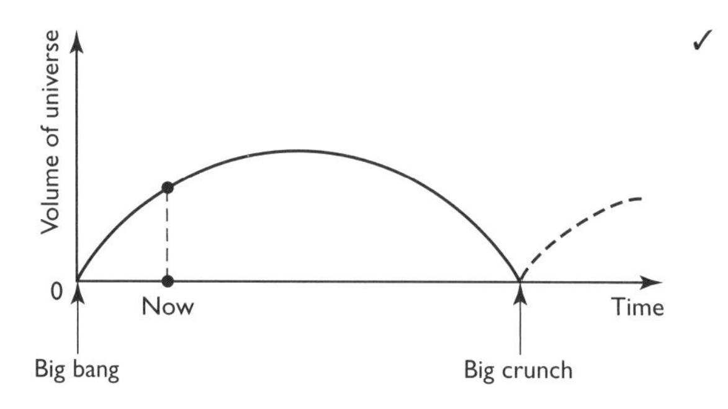

For a closed universe, the density has to be larger than the critical density. ✓

It will stop expanding; it will then contract towards the big crunch. ✓

Candidate B

(c) (ii) In a closed universe, the expansion will eventually stop because of strong gravitational forces and then it will start to contract. ✓

Candidate A has once again produced a superb answer and the diagram is fully labelled. Candidate B has a tendency not to scrutinise the question; there is no labelled diagram. This is shoddy work.

Candidate A has given a model answer and has scored full marks. All the necessary detail is given and the candidate clearly understands the work thoroughly. The answer for part (c) (i) is particularly worthy of praise because it requires extended writing. This candidate is definitely on target to get an A* grade. Candidate B has lost marks because most of the questions were not read properly; missing the graph in part (c) (ii) is just sheer laziness. In addition, the description in part (c) (ii) was too brief for A2.